高等学校计算机专业规划教材

Android程序设计教程
（第二版）

肖云鹏 刘红 刘宴兵 编著

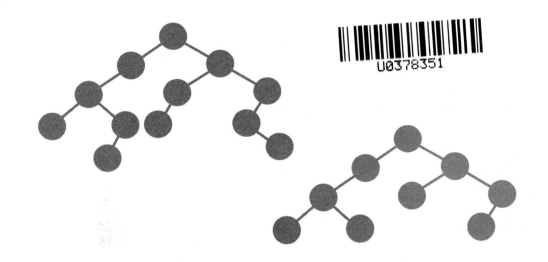

清华大学出版社
北京

内 容 简 介

本书是为大学本、专科 Android 学习准备的教材。全书以 what、why、how 的方式讲解，强调原理，重视实践。以大学期间最常使用教学案例"图书管理系统"贯穿每个知识点，从第 2 章开始，安排在每一章的最后一节。纵向方面，各章承前启后，层层递进，从最简单的单界面、静态数据的图书管理系统→多界面、静态数据的图书管理系统→带本地存储的图书管理系统→带网络连接的图书管理系统→带多媒体的图书管理系统→用 Service 实现新书上架、带异步刷新的进阶功能的图书管理系统。横向方面，每章最后一节的实例也是对该章学习内容的总结和实践。同时，根据实际教学情况，我们在本书的实例中用最简单的方式融汇了已在面向对象、数据结构、数据库、网络编程、多线程、通信协议、程序结构、常用设计模式等部分学习过，但不一定能够实际运用重要知识点。为了配合教师教学及学生自学，本书提供了配套教学的 PPT 和所有章节的源代码，扫描前言中的二维码即可下载。

本书封面贴有清华大学出版社防伪标签，无标签者不得销售。
版权所有，侵权必究。举报：010-62782989，beiqinquan@tup.tsinghua.edu.cn。

图书在版编目（CIP）数据

Android 程序设计教程/肖云鹏，刘红，刘宴兵编著.—2 版.—北京：清华大学出版社，2019.10
（2023.8重印）
（高等学校计算机专业规划教材）
ISBN 978-7-302-51441-1

Ⅰ.①A… Ⅱ.①肖… ②刘… ③刘… Ⅲ.①移动终端-应用程序-程序设计-高等学校-教材
Ⅳ.①TN929.53

中国版本图书馆 CIP 数据核字（2018）第 242358 号

责任编辑：贾 斌 李 晔
封面设计：何风霞
责任校对：梁 毅
责任印制：杨 艳

出版发行：清华大学出版社
　　网　　址：http://www.tup.com.cn, http://www.wqbook.com
　　地　　址：北京清华大学学研大厦 A 座　　邮　编：100084
　　社 总 机：010-83470000　　　　　　　　　邮　购：010-62786544
　　投稿与读者服务：010-62776969, c-service@tup.tsinghua.edu.cn
　　质 量 反 馈：010-62772015, zhiliang@tup.tsinghua.edu.cn
印 装 者：北京国马印刷厂
经　　销：全国新华书店
开　　本：185mm×260mm　　印　张：16.25　　字　数：393 千字
版　　次：2013 年 4 月第 1 版　2019 年 10 月第 2 版　印　次：2023 年 8 月第 5 次印刷
印　　数：6001～7000
定　　价：49.00 元

产品编号：073420-01

本书是为大学本、专科 Android 学习准备的教材。自本书第一版出版以来，有许多高校和同学一直在使用，这给了我很大的激励。本书的第二版，除了更新了第一版的一些表达问题和笔误外，还增加了 Android Studio 的使用以及如何导入本书的案例库代码。笔者总结了多年来教学和工程经验，力争使本书做到：

- 在每一个重要的知识点上，以 what、why、how 的方式讲解。在讲是什么（what）问题的时候，多打比方、多讲故事、多画图。让同学们首先有感性认识，再落实到程序代码层面，让学习的过程从感性认识到理性认识再到量化实现。在讲原理（why）的时候，尽量深入透彻，这是对于同学非常重要的要求。我常和学生们说：只有清楚原理才能做出优秀的程序。最后落实到 how 的问题，即使用的问题。
- 本书对学习者前期基础知识的要求是只要学过一点 Java 语言，能看懂 class，会写 helloworld 程序，就能够学习这本教材。教材里用到的所有示例都尽可能做到内容简单、教学目标明确。
- 全书贯穿一个实例——大学教学最常使用的"图书管理系统"，这个实例从第 2 章开始，安排在每一章的最后一节。纵向方面，各章承前启后、层层递进，从最简单的单界面、静态数据的图书管理系统→多界面、静态数据的图书管理系统→带本地存储的图书管理系统→带网络连接的图书管理系统→带多媒体的图书管理系统→用 Service 实现新书上架、带异步刷新的进阶功能的图书管理系统。横向方面，每章最后一节的实例也是对该章学习内容的总结和实践。
- 根据多年来的教学经验，针对教学中实际存在的问题，我们在本书的实例中用最简单的方式融汇了已在面向对象、数据结构、数据库、网络编程、多线程、通信协议、程序结构、常用设计模式等部分前期课程学习过，但不一定能够实际运用重要知识点。我们的初衷是希望本书不仅仅是一本 Android 程序的教材，更希望通过对这本书的学习，激发同学们的学习热情。如在第 5 章，我们首先从最简单的键盘、显示器 IO 开始，逐步讲到联网，大家会发现在建立网络连接后的数据传输和本地 IO 是一样的。在代码方面，不仅给出了 Android 客户端源代码，还给出了服务器端 Socket Server 和 Web Servlet 的源码、源码分析和数据库脚本。这样做的目的是尽可能深入浅出、融会贯通，同时保证大部分接近零基础的

- 本书作为本科教材，因此这不是一本篇幅很大、面面俱到的介绍Android的书。我们认为本科教学应该是启发式的教学。我们在课堂上常要求同学们大学期间在专业课学习上做到三点：

  （1）扎实的专业基础知识；

  （2）良好的英文读写水平；

  （3）快速掌握陌生知识的能力。

课堂上授课时间有限，学期内有限的课程学习，打好基础，掌握学习方法，相信有兴趣的同学自然会"自学成才"，我想这也是大学学习的要领。正是基于这个想法，本书讲到的都是最重要、最基础的问题，因此在书中没有要求Android SDK版本问题。

- 为了配合教学和同学们自学，本书提供了配套教学的PPT和所有章节的源代码。读者可以扫描下面的二维码下载。

扫码下载完整代码及配套PPT

本书的写作得到了清华大学出版社的支持和帮助，得到了很多宝贵意见。我的学生戴天骥和左佳参与了第二版的修改工作。

最后，感谢我的家人对我工作的支持，感谢实验室的前辈、同事对我工作一直的支持，感谢实验室的年轻人和我一起讨论、拼搏的美好时光。

本书的完成得到重庆市重点研发项目(No.cstc2017zdcyzdyfx0002，cstc2017zdcyzdyfx0092)、重庆市基础科学与前沿技术研究项目(No.cstc2017jcyjAX0099)和重庆市研究生教改项目(No. yjg183081)资助。

编　者

2018年10月

目录

第1章 概述 /1

1.1 搭建环境 ···1
   1.1.1 安装 Android Studio ··1
   1.1.2 安装 SDK ··2
1.2 创建第一个 Android 程序 ···3
   1.2.1 使用 Android Studio 创建一个 Android 工程 ········3
   1.2.2 在模拟器上运行 ··6
   1.2.3 在手机上运行 ··8
1.3 如何导入本书案例库 ··8
1.4 Android 应用程序的构成 ··10
1.5 Android 四大组件 ···13
   1.5.1 活动 ··13
   1.5.2 服务 ··14
   1.5.3 内容提供者 ··14
   1.5.4 广播接收者 ··15
1.6 养成良好的学习习惯 ··15
本章小结 ··16

第2章 活动 /17

2.1 Activity 概述 ··17
   2.1.1 Activity 是什么 ··17
   2.1.2 Activity 生命周期 ··19
   2.1.3 Activity 生命周期的示例 ······································22
2.2 一个 Android 工程的整体结构 ···24
   2.2.1 Android 程序中各种文件夹及文件 ························25
   2.2.2 res 文件夹 ···28
   2.2.3 AndroidManifest.xml 文件 ·····································32
2.3 最简单的图书管理系统 ··33

第 3 章 用户界面 /37

- 3.1 用户界面基础知识 ··37
- 3.2 界面基本组件 ··38
  - 3.2.1 界面基本属性 ···38
  - 3.2.2 TextView ···38
  - 3.2.3 EditText ··40
  - 3.2.4 Button ···42
  - 3.2.5 复选框（CheckBox）···44
  - 3.2.6 单选按钮 ··45
  - 3.2.7 Listview ··47
- 3.3 布局 ···48
  - 3.3.1 FrameLayout（帧布局）···48
  - 3.3.2 LinearLayout（线性布局）··50
  - 3.3.3 RelativeLayout（相对布局）···52
  - 3.3.4 TableLayout（表格布局）···54
  - 3.3.5 AbsoluteLayout（绝对布局）··55
  - 3.3.6 多种布局混合使用 ···56
- 3.4 菜单 ···58
  - 3.4.1 选项菜单 ··58
  - 3.4.2 上下文菜单 ··59
  - 3.4.3 子菜单 ···61
  - 3.4.4 定义 XML 菜单文件 ··63
- 3.5 事件响应 ··65
  - 3.5.1 基本事件 ··65
  - 3.5.2 事件的响应 ··66
- 3.6 界面切换与数据传递 ···68
  - 3.6.1 Intent 与 Bundle ··68
  - 3.6.2 界面切换 ··70
  - 3.6.3 传递数据 ··71
- 3.7 Activity 界面刷新 ···75
- 3.8 Activity 栈及 4 种启动模式 ···75
  - 3.8.1 Activity 栈概述 ···75
  - 3.8.2 Activity 启动模式定义方法 ··76
  - 3.8.3 standard 启动模式 ···77

	3.8.4	singleTop 启动模式	79
	3.8.5	singleTask 启动模式	80
3.9	有多个界面的单机版图书管理系统		83

第 4 章 数据存储 /108

4.1	Preference 存储方式		108
	4.1.1	SharedPreferences	109
	4.1.2	PreferenceActivity	110
	4.1.3	XML 解析	114
4.2	文件的存储		117
	4.2.1	内部存储	117
	4.2.2	外部存储	118
4.3	SQLite 数据库		119
	4.3.1	SQLite 简介	119
	4.3.2	SQLite 数据库基本数据操作	122
	4.3.3	SQLiteOpenHelper 类	124
	4.3.4	数据库文件存储位置（SD 卡/手机内存）	126
4.4	数据共享 ContentProvider		127
	4.4.1	Android 系统自带的 ContentProvider	127
	4.4.2	自定义 ContentProvider	128
4.5	一个有本地数据库的单机版图书管理系统		129

第 5 章 网络编程 /138

5.1	什么是网络编程		138
	5.1.1	Socket 通信	139
	5.1.2	HTTP 通信	140
5.2	客户/服务器模式		140
	5.2.1	控制台上的简单输入输出	141
	5.2.2	控制台上的循环输入输出	141
	5.2.3	一个客户端和一个服务器端一次通信	143
	5.2.4	一个客户端和一个服务器端多次通信	146
	5.2.5	多个客户端和一个服务器端串行通信	151
	5.2.6	多个客户端和一个服务器端并行通信	155
	5.2.7	客户端与服务器端 HTTP 通信	158
5.3	通信协议		161

5.3.1 什么是协议，为什么需要协议 ······ 161
5.3.2 如何实现协议 ······ 161
5.4 Handler 机制 ······ 162
5.5 联网的图书管理系统 ······ 162
  5.5.1 定义协议 ······ 163
  5.5.2 使用 TCP Socket 的图书管理系统 ······ 164
  5.5.3 使用 TCP Socket 的图书管理系统的服务器 ······ 175
  5.5.4 使用 HTTP 的图书管理系统 ······ 194
  5.5.5 使用 HTTP 的图书管理系统的服务器 ······ 195

第 6 章　多媒体　/197

6.1 MediaPlayer ······ 197
6.2 音频播放 ······ 200
  6.2.1 从源文件播放音频 ······ 200
  6.2.2 从文件系统播放音频 ······ 201
  6.2.3 从流媒体播放音频 ······ 203
6.3 视频播放 ······ 204
  6.3.1 从源文件播放视频 ······ 204
  6.3.2 从文件系统播放视频 ······ 204
  6.3.3 从流媒体播放视频 ······ 205
6.4 为图书管理系统配上音乐 ······ 205

第 7 章　图书管理系统程序进阶　/209

7.1 Service（服务） ······ 209
  7.1.1 了解 Service ······ 209
  7.1.2 Service 的启动与生命周期 ······ 209
7.2 系统服务 ······ 216
  7.2.1 什么是系统服务 ······ 216
  7.2.2 获得系统服务 ······ 216
  7.2.3 重力感应 ······ 217
7.3 广播 ······ 218
  7.3.1 什么是广播 ······ 218
  7.3.2 广播的接收与响应 ······ 219
  7.3.3 广播的发送 ······ 220
7.4 Service 实现新书上架通知 ······ 221

          7.4.1 客户端 ·· 221
          7.4.2 服务器 ·· 226
    7.5 带异步刷新功能的图书管理系统 ·· 233
          7.5.1 Tab 标签的实现 ·· 235
          7.5.2 自定义的 ListView 与 Adapter ··································· 237
          7.5.3 异步刷新实现 ·· 241
          7.5.4 其他部分实现 ·· 242

参考文献　/247

第 1 章 概　　述

本章讲述 Android 开发环境的搭建和如何使用 Android Studio 建立一个 Android 工程，以及对 Android 应用程序的文件夹结构及其组件的讲解。

1.1　搭　建　环　境

如果从事 Android 应用程序的开发，好的开发工具和开发环境是必不可少的。安装环境的配置需要如下工具和开发包：JDK、Android Studio 和 Android SDK。

现在来具体讲一下如何获得这些工具和开发包以及它们的正确安装和配置。考虑到读者已有一定的 Java 基础，故在此不会对 Java 的运行环境即 JDK 的安装过程做出讲解。

1.1.1　安装 Android Studio

在安装了 JDK 的基础上来安装 Android Studio。Android Studio 的下载网站是 https://developer.android.google.cn/studio/index.html。这里以 Windows 操作系统为例来讲述如何安装配置 Android 开发环境。

（1）打开网址，单击如图 1-1 所示的下载按钮。

图 1-1　Android Studio 下载页面

（2）根据提示下载完成 Android Studio 之后，启动下载的 .exe 文件，见图 1-2。

图 1-2 Android Studio 安装文件

（3）根据安装向导的指示安装 Android Studio 和所需的 SDK 工具。

在有些 Windows 系统中，启动器脚本无法找到 JDK 的安装位置。如果遇到此问题，即需要设置指示正确位置的环境变量。选择 Start→Computer→System Properties→Advanced System Properties 命令，然后打开 Advanced 选项卡，选择 Environment Variables 选项，添加指向 JDK 文件夹位置（例如 C:\Program Files\Java\jdk1.8.0_77）的新系统变量 JAVA_HOME。

1.1.2 安装 SDK

在安装 Android Studio 的过程中，我们已经安装了 SDK。如果在安装过程中没有安装完所需要的 SDK，在 Android Studio 中应该怎样去安装 SDK？下面在 Android Studio 中安装 SDK。

（1）启动 Android Studio，单击 Android Studio 上的 按钮。

（2）选择所需要的 SDK，如图 1-3 所示。

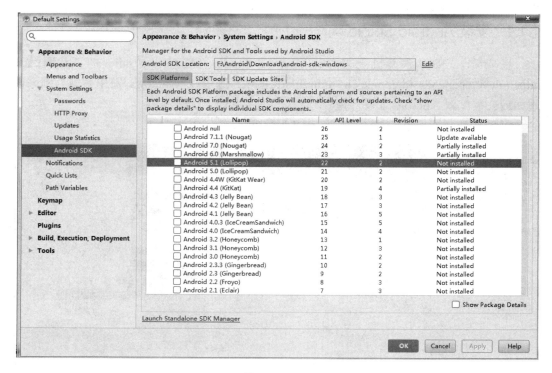

图 1-3 SDK

（3）单击 OK 按钮，等待安装界面如图 1-4 所示，更新完成单击 Finish 按钮。

第 1 章 概　　述

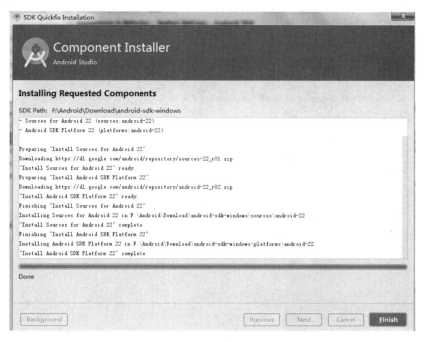

图 1-4　SDK 安装完成

1.2　创建第一个 Android 程序

本节将新建一个简单的 Android 程序来测试 Android Studio 是否安装成功，以及在模拟器和手机上运行该程序。

1.2.1　使用 Android Studio 创建一个 Android 工程

（1）启动 Android Studio，选择 File→New→New Project 命令，如图 1-5 所示。

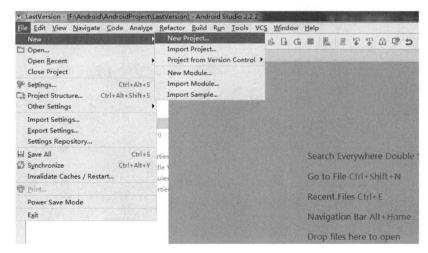

图 1-5　建立 Android 工程

（2）在弹出的对话框中输入工程信息，如图 1-6 所示。

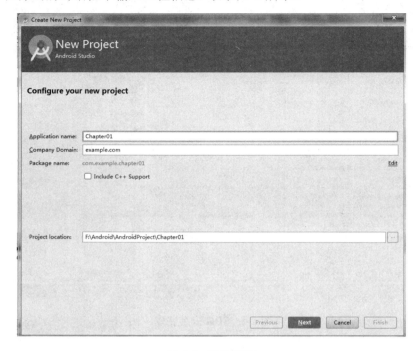

图 1-6　工程信息

（3）单击 Next 按钮，在弹出的对话框中根据业务选择相应的 SDK 版本信息，如图 1-7 所示。

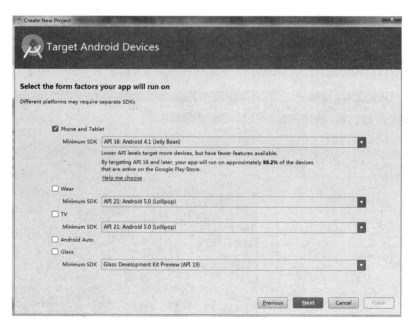

图 1-7　SDK 版本信息

（4）单击 Next 按钮，在弹出的对话框中选择相应的 Activity（Empty Activity 是 Android 应用中最简单的 Activity），如图 1-8 所示。

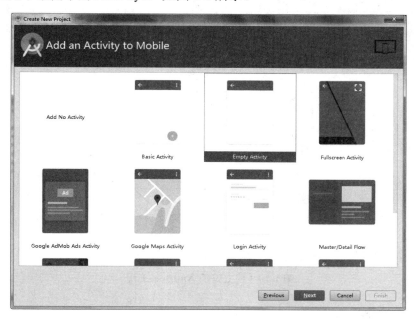

图 1-8 Activity 类型

（5）单击 Next 按钮，在弹出的对话框中输入 Activity 的信息，如图 1-9 所示。

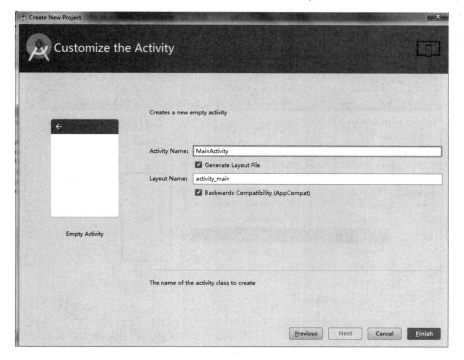

图 1-9 Activity 的信息

（6）单击 Finish 按钮，可得到如图 1-10 所示的界面。

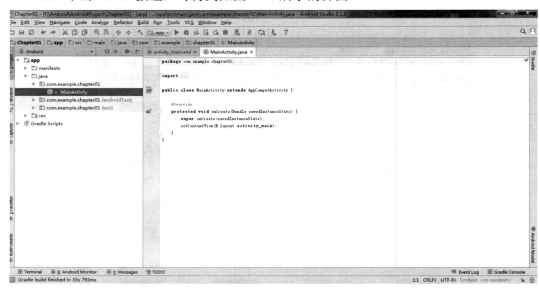

图 1-10　建立好的 Android 工程

1.2.2　在模拟器上运行

在运行 Chapter01 工程之前还需要建立一个 AVD（Android Virtual Device）设备。一个 AVD 设备对应一个 Android 版本的模拟器实例。

（1）单击 Android Studio 上的 按钮。在显示的对话框中单击 Create 按钮新建一个 AVD 设备。选择相应的 AVD，如图 1-11 所示。

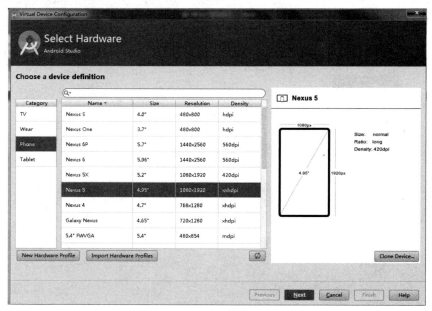

图 1-11　建立 AVD 设备

（2）输入 AVD Name，如图 1-12 所示。

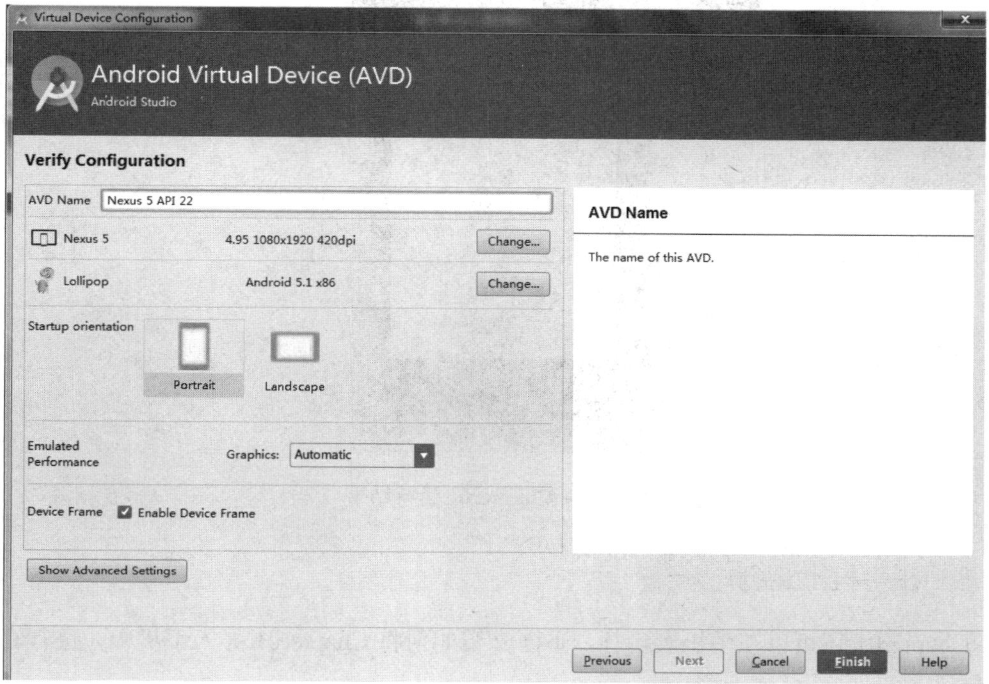

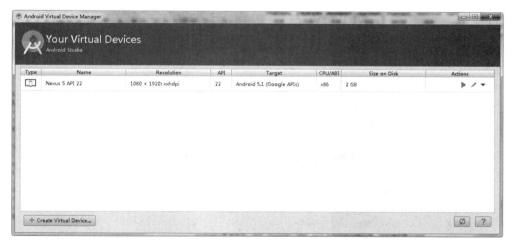

图 1-12　输入 AVD Name

（3）单击 Finish 按钮，完成 AVD 的建立，如图 1-13 所示。

图 1-13　建立好的 AVD 设备

（4）建立完 AVD 设备后。找到 Chapter01 工程，单击上方的 ▶ 按钮，选择 Available Virtual Devices 下创建好的 AVD 设备，运行 Chapter01。这时模拟器会自行启动并运行 Chapter01，可以看到如图 1-14 所示的 Chapter01 运行结果。

图 1-14　Chapter01 运行结果

1.2.3　在手机上运行

Chapter01 还可以在手机上运行，下面介绍如何将 Chapter01 导入手机中，进行真机测试。

打开 Android Studio，将手机与计算机通过 USB 连接，找到 Chapter01 工程，单击上方的 ▶ 按钮，选择 Connected Devices 下已连接的手机，运行 Chapter01 即可。

1.3　如何导入本书案例库

在本书的前言中，已经向读者提供了网址可以下载所有章节的源代码，现在以 Chapter2.3 为例，介绍 Android Studio 中导入案例库。

（1）启动 Android Studio，选择 File→New→Import Project 命令，如图 1-15 所示。

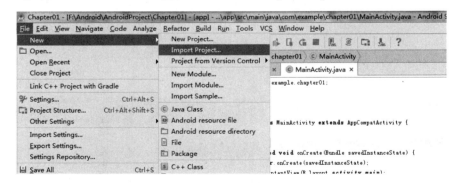

图 1-15　Import Project 导入

（2）选择需要导入的案例 Chapter2.3，如图 1-16 所示。

（3）单击 OK 按钮，选择导入的目标文件夹，如图 1-17 所示。

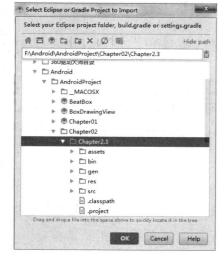

图 1-16　选择 Chapter2.3

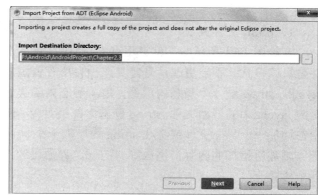

图 1-17　选择目标文件夹

（4）单击 Next 按钮，选择替代选项，最后一个驼峰式命名的选项可以去掉，如图 1-18 所示。

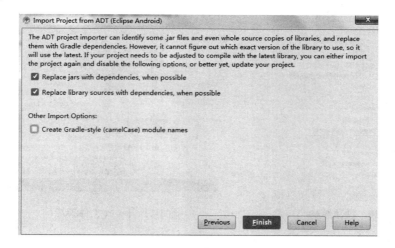

图 1-18　选择替代选项

（5）单击 Finish 按钮，完成 Chapter2.3 案例的导入，结果如图 1-19 所示。

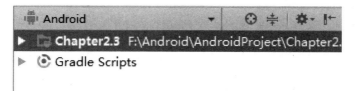

图 1-19　Chapter2.3 案例导入成功

1.4 Android 应用程序的构成

在 Android Studio 中建立的 Android Project 名称为 Chapter01，项目结构如图 1-20 所示。

1．build 文件夹

在如图 1-21 所示的文件夹下有 R.java。R.java 是在建立项目时自动生成的，这个文件是只读模式，不能更改，是定义该项目所有资源的索引文件。默认有 attr、drawable、layout、string 这 4 个静态内部类，每个静态内部类对应一种资源，如 layout 静态内部类对应 layout 中的界面文件，string 静态内部类对应 string 内部的 string 标签。如果在 layout 中在增加一个界面文件或者在 string 内增加一个 string 标签，R.java 会自动在其对应的内部类增加所增加的内容，生成唯一的 id，界面可通过 R 文件来访问工程 res 文件夹中的各个资源。

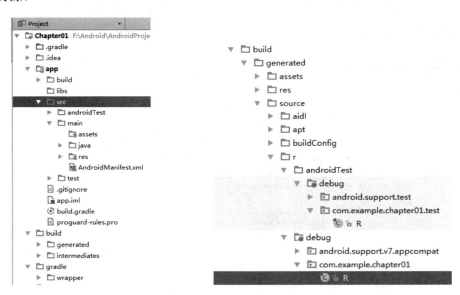

图 1-20　项目结构　　　　　　图 1-21　build 文件夹结构

2．src 文件夹

在如图 1-22 所示的文件夹下包含项目的所有源文件，包括 Java 源文件（main\java 文件夹下）、各种资源文件等。

3．assets 文件夹

此文件夹是资产文件夹，该文件夹下的文件不会被映射到 R.java 中，访问的时候需要 AssetManager 类以字节流的方式来读取。

4．res 文件夹

是资源文件夹，可以存放一些图标、界面文件和应用中用到的文字信息，如图 1-23 所示。

第 1 章 概　述

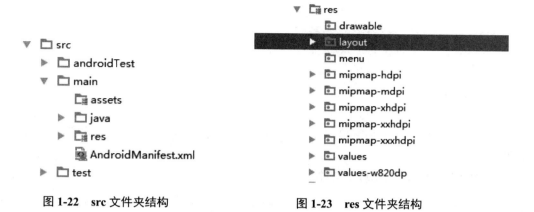

图 1-22　src 文件夹结构　　　　图 1-23　res 文件夹结构

5. mipmap-*dpi 文件夹

将图标按分辨率的高低放入不同的文件夹，其中 mipmap -hdpi 用来存放高分辨率的图标，mipmap-mdpi 用来存放中等分辨率的图标，mipmap-ldpi 用来存放低分辨率的图标，mipmap -xhdpi 用来存放超高分辨率的图标。

6. layout 文件夹

此文件夹用来存放界面信息。本例中的布局文件是自动生成的 activity_main.xml，如下代码：

```
1.<RelativeLayout xmlns:android="http://schemas.android.com/apk/res/android"
2.    xmlns:tools="http://schemas.android.com/tools"
3.    android:layout_width="match_parent"
4.    android:layout_height="match_parent" >
5.
6.    <TextView
7.        android:layout_width="wrap_content"
8.        android:layout_height="wrap_content"
9.        android:layout_centerHorizontal="true"
10.       android:layout_centerVertical="true"
11.       android:text="@string/hello_world"
12.       tools:context=".MainActivity" />
13
14.</RelativeLayout>
```

<RelativeLayout>是相对布局的标签，<TextView>标签代表一个 TextView 组件，在 TextView 组件中显示输入的文字。

7. menu 文件夹

此文件夹存放 Android 菜单的布局文件，本例中的菜单文件是自动生成的，代码如下：

```
1.<menu xmlns:android="http://schemas.android.com/apk/res/android">
2.    <item android:id="@+id/menu_settings"
```

```
3.        android:title="@string/menu_settings"
4.        android:orderInCategory="100"
5.        android:showAsAction="never" />
6.</menu>
```

8. values 文件夹

此文件夹存放储存值的文件。该文件夹下的 strings.xml:用来定义字符串和数值，每个 string 标签声明了一个字符串，name 属性指定它的引用值。styles.xml 是定义样式（style）对象。

9. AndroidManifest.xml 文件

```
1.<manifest xmlns:android="http://schemas.android.com/apk/res/android"
2.    package="ui.cqupt"
3.    android:versionCode="1"
4.    android:versionName="1.0" >
5.
6.    <uses-sdk
7.      android:minSdkVersion="8"
8.      android:targetSdkVersion="15" />
9.
10.   <application
11.       android:icon="@drawable/ic_launcher"
12.       android:label="@string/app_name"
13.       android:theme="@style/AppTheme" >
14.       <activity
15.           android:name=".MainActivity"
16.           android:label="@string/title_activity_main" >
17.           <intent-filter>
18.               <action android:name="android.intent.action.MAIN" />
19.
20.               <category android:name="android.intent.category.LAUNCHER" />
21.           </intent-filter>
22.       </activity>
23.   </application>
24.
```

每一个 Android 工程都有一个名为 AndroidManifest.xml 的配置文件，在所有项目中该文件的名称不变。该文件是 Android 工程的一个全局配置文件，所有在 Android 中使用的组件（如 Activity、Service、ContentProvide 和 Broadcast Receiver）都要在该文件中声明，并且该文件还可以声明一些权限以及 SDK 的版本等信息。

该文件第一行是 XML 文件的版本和编码的声明。下面是 manifest 根元素，该元素中指定了命名空间、包名称、版本代码号和版本名称等信息。

Application 子元素中的 3 个属性分别指定程序的图标、标题和主题。下面是 Activity

组件的声明，Activity 有两个属性表明 Activity 类的名称和标题。

<intent-filter>是找到该 Activity 的过滤器，这里的 action 表示该 Activity 是程序的入口。category 表明在加载程序时运行。

<uses-sdk>的两个属性表明使用的 SDK 的最低版本和当前版本。

1.5　Android 四大组件

在 1.3 节已经对 Android 应用程序的文件夹结构以及其中每个文件的功能做出了大致介绍，想要进行应用程序开发，还需要对 Android 应用程序的构造有更深一层的理解。Android 应用程序没有使用常见的应用程序入口点的方法（例如 main()方法），取而代之的是一系列组件。Android 应用程序是由组件组成的，而这些组件是可以相互调用、相互协调、相互独立的基本功能模块。一般情况下，一个 Android 应用程序是由以下 4 种组件构成的：活动（Activity）、服务（Service）、内容提供器（Content provider）和广播接收者(Broadcast receiver)，它们合称为"Android 的四大组件"，如表 1-1 所示。本节主要介绍这 4 种组件的概念。

表 1-1　四大组件的定义

组件名称	组件类/接口	定　义
Activity	Android.app.activity	与用户进行交互的可视化界面（屏幕），类似窗体的组件
Service	Android.app.service	无界面的、生命周期长的、运行在后台、关注后台事务的组件
Content provider	Android.contentprovider	实现不同的应用程序之间数据共享的组件
Broadcast receiver	android.content. BroadcastReceiver	接收并且响应广播消息的组件

1.5.1　活动

活动（Activity）是 Android 应用程序最基本的组件，显示可视化的用户界面，是用户和应用程序交互的窗口，并接收与用户交互所产生的界面事件。在 Android 程序中一个 Activity 代表了一个单独的屏幕，好比 ASP.NET 应用中的一个网页界面或者 C/S 程序中的窗体（Form）。在 Activity 的这个可视化区域的屏幕中可以添加 View，并对 View 做一些操作，View 可以理解为一个 UI[即 User Interface（用户界面）的简称]容器。在这个容器中有很多 UI 元素供开发者添加到屏幕上，比如，Button、TextView、EditView、List 等，这些丰富的 UI 元素组成了和用户交互时的丰富的用户界面。

Android 应用程序可以包含一个或多个 Activity，一般在程序启动以后会呈现一个 Activity，用于提示用户程序已经正常启动。Activity 在界面上的表现形式一般是全屏窗体或非全屏的悬浮窗体或对话框。

我们可以实现从一个屏幕切换到另一个屏幕，并且完成新的活动。我们知道，Android 会把每个应用从开始到当前的每一个屏幕的页面都压入到堆栈中，当打开一个新的屏幕时原来的屏幕会被置为暂停状态，并且压入到历史的堆栈中。可以通过返回操作来弹出栈顶的屏幕并设置为当前操作的屏幕界面，也可以有选择地移除一些堆栈中不会用到的屏幕，即关掉不需要的界面。

下面以一个简单的示例说明一下活动（Activity）：假如一个服装设计师要在白纸上画设计图，他可以在上面任意添加图形，这个图形也就类似于这里讲的 view 中的元素，一个设计师想要设计出令人满意的服装设计图，那他就要不停地在设计纸上画图设计，一张又一张，令人满意的就留下来存在那里，不满意的就扔掉。这就好似我们的 Activity 窗体，可以有很多个，而且存在栈中，不想要这个 Activity 就随时删除（关掉）。

一个 Activity 就是一个独立的类（是 Android 的核心类，该类的全名是 android.app.Activity），继承活动的基类而来。

在开发过程中，Activity 是由 Android 系统进行维护的，它有自己的生命周期。

Activity 生命活动周期，即"产生、运行、销毁"，但是这其中会调用许多方法，如 onCreate（创建）、onStart（激活）、onResume（恢复）、onPause（暂停）、onStop（停止）、onDestroy（销毁）、onRestart（重启）。这些将在后面章节讲到。

1.5.2 服务

Android 中的服务（Service）：类似于 Windows 系统中的 Windows Service，它运行在后台，是不可见的、没有界面的、生命周期长的组件。一些常见的例子包括：手机 QQ 程序，它可以在转到后台运行的时候保持接收信息；媒体播放器程序，它可以在转到后台运行的时候保持播放歌曲；或者文件下载程序，它可以在后台执行文件的下载等。

下面以一个简单的示例说明一下服务（Service）：

手机 QQ 应该有很多 Activity。当登录手机 QQ 成功时会看到一个主界面（屏幕）即 Activity，里面有很多列表（好友列表、陌生人列表、企业列表等等），我们可以切换到好友列表界面并选择一个好友切换到聊天界面进行聊天。然而，当我们想要浏览其他网页或者打开音乐界面时，手机 QQ 就通过启动一个 Service，从而使手机 QQ 在后台运行，虽然没有界面，但是它并没有退出程序，而是一直运行在后台，当有新的消息发过来时，可以单击之后马上回到聊天界面，这就是由服务来保证当用户界面关闭时，仍然能接收到消息。Service 具体的功能作用以及使用方式将在以后章节详细讲到，这里只做一下大致了解即可。

1.5.3 内容提供者

内容提供者（ContentProvider）是 Android 系统提供的一种标准的共享数据的机制，应用程序可通过 ContentProvider 访问其他应用程序的私有数据。私有数据可以是存储在文件系统中的文件，也可以是 SQlite 中的数据库。Android 系统内部也提供一些内置的 ContentProvider，能够为应用程序提供重要的数据信息。

Android 系统有一个独特之处：数据库只能被它的创建者所使用，也就是说，数据是私有的，其他的应用是不能访问的，但是如果一个应用要使用另一个应用的数据（即：不同应用间的数据共享）该怎么做呢？那么这个时候 ContentProvider 就派上用场了，它是一种特殊的存储数据的类型，一个 ContentProvider 提供了一套标准的方法接口用来获取以及操作数据，能使其他应用程序保存和读取此 ContentProvider 提供的各种数据（这些 ContentProvider 数据类型包括音频、视频、图片以及私人通讯录等）。那么怎么来实现数据共享呢？只要实现 ContnentPriviver 的接口就可以了。

ContentPrivider 已经实现了数据的封装和处理，外界是看不到数据的具体存储细节的，只需要通过这些标准的接口打交道就可以了，它可以实现读取数据、删除数据、插入数据等操作。

总的来说，ContentProvider 的主要作用如下：

（1）为存储和读取数据提供了统一的接口。

（2）使用 ContentProvider，应用程序可以实现数据共享。

（3）Android 内置的许多数据都是使用 ContentProvider 形式，供开发者调用的（如视频、音频、图片、通讯录等）。

1.5.4 广播接收者

广播接收者（BroadcastReceiver）是用来接收并响应广播消息的组件，跟 Service 一样是不可见的，没有用户界面，它的唯一作用就是接收并响应消息。但它可以通过启动 Activity 或者 Notification 通知用户接收到重要信息（Notification 能够通过多种方法提示用户，包括闪动背景灯、振动设备、发出声音或在状态栏上放置一个持久的图标等）。

很多时候，广播消息是由系统发出的。例如，电池的电量不足、未接电话显示图标、收到短信等。除此之外，应用程序还可以发送广播消息。例如，通知其他的程序数据已经下载完毕并且这些数据可以使用了。

一个应用程序可以有多个广播接收者，所有的广播接收者类都需要继承 android.content.BroadcastReceiver 类。

当系统中发送出来一个意图后，系统会根据该意图的 action 自动去匹配系统中现有的各个意图过滤器 Intent-filter，一旦发现有匹配的广播接收者，系统会自动调用该广播接收者的 onReceive 方法，那么就可以通过这个方法做事了。

广播接收者做的事情不宜太复杂或耗时太长，在系统中 BroadcastReceiver 的生命周期约 5s，超时后将会被回收，所以如果需要在 onReceive 方法做复杂的业务处理，最好开启一个线程来完成工作。

1.6 养成良好的学习习惯

在前言部分已经讲过，我们要求同学们大学期间在专业课学习上做到 3 点：

(1) 扎实的专业基础知识；

(2) 良好的英文读写水平；

(3) 快速掌握陌生知识的能力。

本书的目标是作为本科教材，因此这不是一本很厚的、面面俱到的 Android 书，而且我们认为本科的教学本身也应该是启发式的教学。对于 Android 的学习，也是一样，一定要掌握原理，理论性知识和上机实践同样重要。

到目前为止，应用好以下几样资源：书、Android Studio 和 API 文档（按照本章上面介绍的安装方法，本书的 Android API Doc 在 F:\android-sdk-windows\docs\reference\packages.html 路径下）和网络（应用好 baidu、google）。

本 章 小 结

通过上述介绍，我们基本了解了 Android 环境的搭建与配置、一个简单的 Android 应用程序的创建及其基本结构和运行机制、Android 的主要组件的概念。

第 2 章 活 动

在第 1 章的学习中我们提到了，Android 系统由 Activity、Service、Broadcast、Receiver 和 ContentProvider 组成。其中 Activity 是使用频率最高、最重要的组件。在 Android 系统中 Activity 提供可视化的用户界面，一个 Android 系统通常由多个 Activity 组成。Activity 有自己的生命周期，由 Android 系统控制。

本章将对 Activity 做详细介绍。

2.1 Activity 概述

2.1.1 Activity 是什么

在 1.4 节已经大致介绍了一下 Activity 的概念，这里首先感性地认识一下 Activity 是什么。

打个比方，图 2-1 讲的是墙上镶嵌了壁橱，壁橱放置了各种各样的东西。墙就类似我们讲的 Activity，壁橱就类似 layout 布局管理器，壁橱上的东西就类似那些 UI 元素，墙的背后或许还有我们看不见的东西，而这些墙背后的东西就是供前端界面调用的非前端界面的 Java 类，这些普通的 Java 类就是我们在学习 Java 语言时学到的那些。

图 2-1 对 Activity 的感性认识

Activity 是什么？其实 Activity 就好比一面墙，我们可以在墙上添加一个壁橱，壁橱的形状、大小、颜色以及放置的位置都由我们自己决定，这个壁橱就好似 Android 程序中的 layout 布局管理器。如图 2-1 所示，我们可以在壁橱上放置任意我们喜欢的东西，各种书籍、花瓶、画、床单、被子等。这些放置在壁橱上的东西就好似前面讲过的 UI 元素。可是这只是表面能看见的东西，还有一些是看不见的，隐藏在这面墙的后面，或许是一个水管，或许是一条电源线路，或许是天然气管道，也许是一根网线等。落实到 Android 程序中，这些隐藏在墙背后我们看不到的东西就是那些 Java 实体类、逻辑控制类、网络连接类等。

不同的墙有不同的壁橱，壁橱有不同的形状，壁橱里放在不同的东西。也许卫生间的墙也有个"壁橱"（哈哈，卫生间里是不是叫这个名字？），里面放的可能就是肥皂、洗发水、淋浴，等等，淋浴后面连的是水管。当我们打开淋浴时，淋浴"调用"了壁橱背后的水管。

在 Android 程序中，通常一个界面对应一个 Activity。实际上，Activity 代表着 Android 设备用户界面显示的那块"区域"。一个 Activity 对应一个 layout（布局），一个 layout 中放置我们设计的用户界面组件，例如按钮、文本框等。

这里有两个问题是在教学中常常遇到的：

（1）必须是一个界面对应一个 Activity 吗？刚才说了，通常一个界面对应一个 Activity。这就好比画画，通常画家一幅画要用一张新的白纸，但是我们也可以在开始新的画的时候把上一张画全部擦掉，再在同一张纸上继续画新的画，但是这样的画家真的是太少了！落实到程序里，就是在一个程序中用一个 Activity，在界面之间切换时换上新的 layout，在新的 layout 上用新的 UI 组件。这样做且不说程序的可读性，起码是不符合基本的面向对象思想。就像我们在 Java 程序中所有的界面用一个 frame，为了实现多界面，不停地换 layout，换 UI component，这样做显然是不合适的。熟悉 Java Web 的同学想想，我们把一个应用所有的逻辑都放在一个 servlet 中实现是不是太让人吃惊了。说得再专业一点，每个 Activity 都是有生命周期的（下面马上就要介绍生命周期），在生命周期的不同阶段应用程序员要复写不同的回调函数。这样，就更不应该多个界面公用一个 Activity，切换不同的 layout，这样回调函数里写什么都不合适。生命周期的问题，读者可能暂时不懂，没关系，2.1.2 节就要讲，学完再回头来看这部分内容，相信会有更进一步的认识。

（2）一个 Activity 对应一个 layout 吗？这个问题的答案是不一定。就像我们刚才举的例子一样，有些时候不同的墙也许用的就是同一个壁橱，只是壁橱里的东西不同而已。在我们学习网页设计的时候也知道，不同网页 html 也许只是内容不同，"风格"是一样的，用的是同一个 css。在 Android 程序中，不同的界面设计也是一样，如果两个界面风格一样，那么就可以用一个 layout。

刚才的例子中说到了墙，墙也是有生命的，我们可以决定它什么时候被创建、什么时候被重新装修、什么时候被销毁等。同样，Activity 也是有生命的。下面详细介绍 Activity 的生命周期——用一个关于生命周期的示例来阐述关于生命周期全部内容。

2.1.2 Activity 生命周期

1．生命周期是什么

什么是生命周期？我们同样引用上面的例子把 Activity 比喻为墙。当主人需要的房屋修建完成的时候墙就被创建了，当然不止一面墙。当主人想在某一面墙上镶嵌一个框架或者放置一扇门的时候，这面墙就被使用，当主人对这面墙的布局不喜欢的时候，可以重置这面墙。当主人觉得这面墙毫无用处的时候，可以把这面墙销毁。这就是墙的整个生命周期，主人就好似那个 Android 系统，决定了墙的创建、调用、销毁。

由此可知，生命周期就是一件物品、软件或者程序从产生到销毁所经历的几个阶段。在大部分情况下，每个 Android 应用程序都将运行在自己的 Linux 进程中。当这个应用的某些代码需要执行时，进程就会被创建，并且将保持运行，直到该进程不再需要，而系统需要释放它所占用的内存，为其他应用所用时才停止。

Android 有一个重要并且特殊的性质就是，一个应用的进程的生命周期不是由应用程序自身直接控制的，而是由系统默认调用，它是根据运行中的应用的一些特征来决定的，包括这些应用程序对用户的重要性、系统的全部可用内存。

2．为什么需要生命周期

为什么需要生命周期呢？其实自然界各种物质都是有生命周期的，人也一样，从生到死。当然，生命周期不止生、死两个状态。就像人一样，我们一辈子、每一阶段都有很多很多状态。之所以引入生命周期，是 Aandroid 平台管理 Activity 的需要。

先打个比方，上幼儿园的小朋友，如果犯了错误，不听老师话，那么老师会打电话给爸爸；如果生病了，那么老师会打电话给妈妈；如果遇到坏人就打 110 等（这只是个比方，方便读者理解，可能不恰当或不适合现代幼儿教学理念）。这里老师就是 Android 系统，负责在孩子的不同状态下通知调用不同的人（接口）。孩子就是各个不同的 Activity，爸爸、妈妈、110 就是系统定义、由系统调用，供程序员重写的接口。至于爸爸做什么、妈妈做什么、110 来了做什么，那是应用程序员的事。

落实到 Android 开发中，比如你正在打着游戏，突然来了电话，这时系统通知你的游戏程序，游戏程序得到通知后要做相应的处理，比如保留当前一些状态变量等，方便用户打完电话回来继续游戏。这就是生命周期的必要性，理解了生命周期，就懂得了系统和应用程序员的接口（默契）。

学过 Java Web 的同学会想到，Servlet 也有生命周期，所以说 Tomcat 是 Servlet 容器。这和 Activity 是类似的。

3．Activity 生命周期

图 2-2 是一张来自官网的 Activity 生命活动周期的图片。

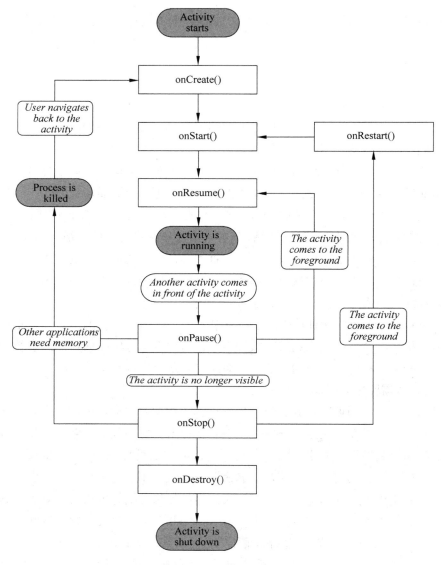

图 2-2　Activity 生命周期

在 Activity 从建立到调回的过程中，需要在不同的阶段调用 7 个生命周期方法。这 7 个方法定义如下：

```
protected void onCreate(Bundle savedInstanceState);
protected void onStart();
protected void onRestart();
protected void onResume();
protected void onPause();
protected void onStop();
protected void onDestroy();
```

Activity 生命周期的事件回调函数如表 2-1（注意：这些函数均由系统调用）所示。

表 2-1 Activity 生命周期事件函数

函　　数	是否可终止	说　　明
onCreate()	否	Activity 启动后第一个被调用的函数,常用来进行 Activity 的初始化，例如创建 View、绑定数据或恢复信息等
onStart()	否	当 Activity 显示在屏幕上时，该函数被调用
onRestart()	否	当 Activity 从停止状态进入活动状态前，调用该函数
onResume()	否	当 Activity 能够与用户交互，接收用户输入时，该函数被调用。此时的 Activity 位于 Activity 栈的栈顶
onPause()	是	当 Activity 进入暂停状态时，该函数被调用。一般用来存放持久的数据或释放占用的资源
onStop()	是	当 Activity 进入停止状态时，该函数被调用
onDestroy()	是	在 Activity 被终止前，即进入非活动状态前，该函数被调用
onSaveInstanceState()	否	Android 系统因资源不足终止 Activity 前调用该函数，用于保存 Activity 的状态信息，供 onRestoreInstanceState() 或 onCreate()恢复之用
onRestoreInstanceState()	否	恢复 onSaveInstanceState()保存的 Activity 状态信息，在 onStart()和 onResume ()之间被调用

由图 2-2 可知，Activity 生命周期是指 Activity 从创建启动到销毁的过程，在这个过程中 Activity 一般表现为 4 种状态：

（1）活动状态。
（2）暂停状态。
（3）停止状态。
（4）非活动状态。

Activity 生命周期中 4 种状态的变换关系如图 2-3 所示。

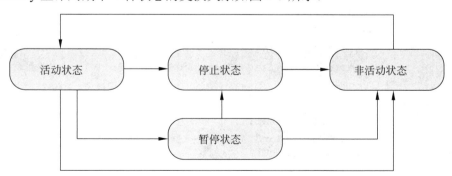

图 2-3 4 种状态

活动状态：当前的 Activity 处于屏幕的前台即使 Activity 在屏幕的最上层，对用户完全可见，并能够与用户交互，这时处于活动状态。

暂停状态：如果一个 Activity 在界面上部分被遮挡，不再处于屏幕最上层，且不能够与用户交互，则这个 Activity 处于暂停状态（Paused）。一个处于暂停状态的 Activity 依然保持活力（保持所有的状态、成员信息，并与窗口管理器保持连接），但是在系统内存极端低下的时候将被杀掉。

停止状态：如果一个 Activity 被另外的 Activity 完全覆盖掉，用户完全看不见，这时 Activity 处于停止状态（Stopped）。它依然保持所有的状态和成员信息，但是它不再可见，所以它的窗口被隐藏，当系统内存需要被用在其他地方的时候，处于停止状态的 Activity 将被杀掉。

非活动状态：前面讲的 3 种状态（活动状态、暂停状态、停止状态）是 Activity 的主要状态，除此之外，Activity 处于非活动状态。如果一个 Activity 是暂停或者停止状态，系统可以将该 Activity 从内存中删除，Android 系统采用两种方式进行删除，要么要求该 Activity 结束，要么直接杀掉它的进程。当该 Activity 再次显示给用户时，它必须重新开始 onStart()和重置 onRestart()前面的状态。

2.1.3 Activity 生命周期的示例

下面通过一个实例来测试 Activity 的生命周期中各个方法的调用情况，这里将覆盖 Activity 中所有的方法，并通过 DDMS 中的 LogCat 来观察。

首先，需要在程序启动的默认的第一个界面中，在最主要的 Java 类中加入一些代码。代码如下：

```
1.  import android.os.Bundle;
2.
3.  import android.util.Log;
4.  import android.app.Activity;
5.
6.  public class  TestActivity extends Activity {
7.
8.      @Override
9.      protected void onCreate(Bundle savedInstanceState) {
10.         Log.d("TAG", "onCreate----------------");
11.         super.onCreate(savedInstanceState);
12.     }
13.
14.     protected void onStart() {
15.         Log.d("TAG", "onStart----------------");
16.         super.onStart();
17.     }
18.
19.     protected void onRestart() {
20.         Log.d("TAG", "onRestart----------------");
21.         super.onRestart();
22.     }
```

```
23.
24.     protected void onResume() {
25.         Log.d("TAG", "onResume----------------");
26.         super.onResume();
27.     }
28.
29.     protected void onPause() {
30.         Log.d("TAG", "onPause----------------");
31.         super.onPause();
32.     }
33.
34.     protected void onStop() {
35.         Log.d("TAG", "onStop----------------");
36.         super.onStop();
37.     }
38.
39.     protected void onDestroy() {
40.         Log.d("TAG", "onDestroy----------------");
41.         super.onDestroy();
42.     }
43. }
```

测试时可以打开 LogCat（通过 Window→Show View→Other→Android→Logcat 命令），得到如图 2-4 所示的窗口，单击 （Create Filter）按钮设置一个过滤器，如图 2-5 所示。

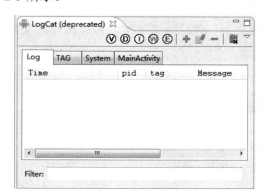

图 2-4　打开的 LogCat

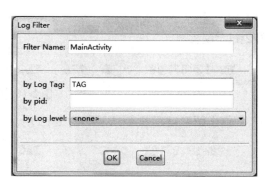

图 2-5　新建 Filter

运行该项目。在 DDMS 中，很容易就能够找到 Log 打印信息。打开应用时先后执行了 onCreate()→onStart()→onResume()这 3 个方法，如图 2-6 所示。

```
07-26 23:28...    D    821    TAG        onCreate----------------
07-26 23:28...    D    821    TAG        onStart----------------
07-26 23:28...    D    821    TAG        onResume----------------
```

图 2-6　启动程序

然后按下返回键，这个程序结束，如图 2-7 所示。

```
07-26 23:28...    D  821   TAG        onPause------------------
07-26 23:28...    D  821   TAG        onStop-------------------
07-26 23:28...    D  821   TAG        onDestroy----------------
```

图 2-7　按返回键退出程序

现在，再回到按返回键之前的状态（即程序正常运行时的状态），此时如果按下 Home 键退出程序，则执行结果如图 2-8 所示。

```
07-26 23:45...    D  1002  TAG        onPause------------------
07-26 23:45...    D  1002  TAG        onStop-------------------
```

图 2-8　按 Home 键退出

由图 2-8 不难知道，这个时候应用程序并没有被摧毁掉（Destory），而是先后执行了 onPause()->onStop() 这两个方法。当再次启动应用程序时，执行结果如图 2-9 所示。

```
07-26 23:55...    D  1002  TAG        onRestart----------------
07-26 23:55...    D  1002  TAG        onStart------------------
07-26 23:55...    D  1002  TAG        onResume-----------------
```

图 2-9　再次启动

此时，如果有电话接入，那么此程序的执行结果如图 2-10 所示。

```
07-26 23:45...    D  1002  TAG        onPause------------------
07-26 23:45...    D  1002  TAG        onStop-------------------
```

图 2-10　电话接入

如果在程序的最初状态，即 OnResume 状态下，此时如果锁屏，那么执行结果如图 2-11 所示。

```
08-13 07:00...    D  776   TAG        onPause------------------
```

图 2-11　锁屏状态

单击解锁键后执行结果如图 2-12 所示。

```
08-13 07:01...    D  776   TAG        onResume-----------------
```

图 2-12　解锁状态

2.2　一个 Android 工程的整体结构

在创建了一个 Android 工程后，可以在左边的文件夹中看见很多文件夹文件，本节将仔细介绍 Android 工程的整体结构，其中 src 和 AndroidManifest 以及 res 是最常用、最基本的，我们会介绍得更加详细一些。src 文件夹放置的是控制 Activity 界面的普通 Android 和 Java 类。res 文件夹是用来保存资源文件的。所谓资源，就是指非代码部分。

例如，我们在 Android 程序中要使用图片来设置桌面，要使用一些字符串来显示信息，那么这些图片、字符串、字体就叫作 Android 中的资源文件，它们中的每一个都会在 gen 文件下的 R.java 中生成对应的 Id 标识符。AndroidMainnfest 是应用程序清单文件，组件必须在此声明后方可用，如图 2-13 中的 Activity1、Activity2、……、Activity*n* 都需要在 AndroidMainfest 中声明。此外，它还声明一些权限信息和其他配置信息。

图 2-13 表明了 Android 工程中各个文件之间的关系，形象地说明了工程的整体结构。

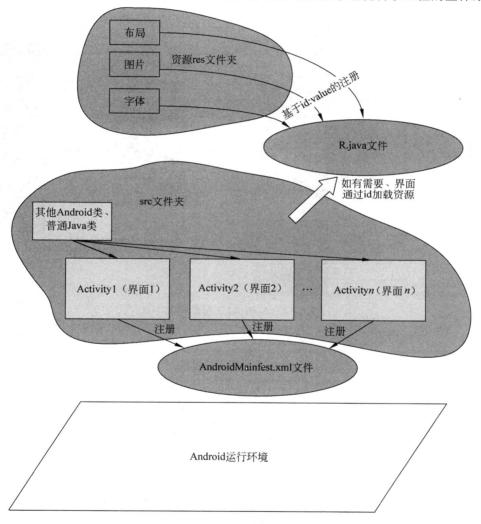

图 2-13　Android 工程整体结构

2.2.1　Android 程序中各种文件夹及文件

第 1 章中已经大致讲过一些有关 Android 程序中的文件夹及文件，看了图 2-13 和图 2-14，相信大家对 Android 的整体结构也会有一个清晰的印象，这里再详细地介绍一下，然后再回顾图 2-13 和图 2-14 就会有很大的收获。当新建一个 Android 程序时，会看到如图 2-15 所示的文件夹。

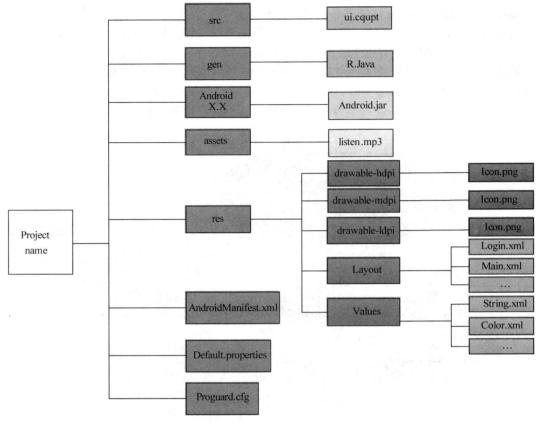

图 2-14　Android 工程结构图

图 2-15　程序的文件夹及文件

src 文件夹：其中是我们写的普通 Java 类，它继承于 Activity 的类，即项目源代码（如果工程比较复杂，可以分成许多包，每个包中可以有很多类，这样可以使代码结构清晰，便于开发人员阅读）。

gen 文件夹：由 ADT 自动生成，其中 R.java 中会生成 res 文件夹中每个文件的 ID，便于在源代码中引用。下面是 R.java 中的代码部分。

```
1.  /* AUTO-GENERATED FILE.  DO NOT MODIFY.
2.   *
3.   This class was automatically generated by the
4.   aapt tool from the resource data it found.  It
5.   should not be modified by hand.
6.   */
7.
8.  package my.system;
9.
10. public final class R {
11. public static final class attr {
12. }
13. public static final class drawable {
14. public static final int ic_launcher=0x7f020000;
15. }
16. public static final class layout {
17. public static final int main=0x7f030000;
18. }
19. public static final class string {
20. public static final int app_name=0x7f040001;
21. public static final int content=0x7f040000;
22. }
23. }
```

Android x.x.x 文件：在新建一个 Android Project 时，选择用于开发的 SDK 版本号。其中 Android.jar 文件封装了绝大部分开发用的工具包，有 j2SE 中的包、Apache 项目中的包，还有 Android 自身的包文件。例如：

android.app——提供高层的程序模型，提供基本的运行环境。

android.database——通过内容提供者浏览和操作数据库。

android.location——定位和相关服务的类。

android.net——提供帮助网络访问的类，超过通常的 java.net.* 接口。

android.os——提供了系统服务、消息传输、IPC 机制。

android.provider——提供类访问 Android 的内容提供者。

android.telephony——提供与拨打电话相关的 API 交互。

android.view——提供基础的用户界面接口框架。

android.widget——各种 UI 元素（大部分是可见的）在应用程序的屏幕中使用。

assets 文件夹——资产文件夹，存放应用程序资源的文件夹，一般放一些较大的文件如字体、视频、音频文件，它们不会被编译，其中的文件不会生成资源文件的 ID，但是会被封装到 apk 中。存放在 res 文件夹中的资源文件，必须通过 ID 来访问。存放在 assets 文件夹中的文件采用传统的地址访问（路径）方式。只读不能写。

bin 文件夹——这个文件夹中存放的是编译之后的应用程序，即 apk 文件。

res 文件夹——存放应用程序用到的资源文件，包含 anim、drawable、layout、values、raw 等文件夹。当这个文件夹下的文件发生变化时，src 文件夹下面的 R.java 就会自动发生变化。res/raw 和 assets 的不同点是，res/raw 中的文件会被映射到 R.java 文件中，访问的时候直接使用资源 ID 即 R.id.filename；assets 文件夹下的文件不会被映射到 R.java 中，访问的时候需要 AssetManager 类。并且 res/raw 不可以有文件夹结构，而 assets 则可以有文件夹结构，也就是在 assets 文件夹下可以再建立文件夹。

AndroidManifest.xml 文件夹——相当于一个部署文件，指明应用所在的包，应用的图标、对 Activity、Service、broadcast 进行声明注册之后，通过 Intent-Filter（意图过滤器）就能实现组件之间的切换，还需设置应用的权限、版本。

proguard.cfg 文件夹——当新建一个 Android 工程之后，会在工程的根文件夹下自动创建一个 proguard.cfg 文件。它是代码混淆工具，通过 proguard，检查并删除没有使用到的类、字段、方法和属性，它优化字节码并去除没有使用到的指令，它使用无意义的名字来重命名使用的类、字段和方法，还可验证代码。别人即使反编译你的 APK 文件，也只能看到很难懂的代码。

project.properties 文件——此文件由 Android Tools 自动生成，它不可以被修改，即使修改了也会被删除，当所在应用程序被安装时，该文件会被安装设备的版本控制系统检测（版本），是否允许安装。

Referenced Libraries 文件夹——外部 jar 包导入文件，如图 2-16 所示，此文件夹可有可无（视应用程序需要）。

Referenced Libraries
commons-collections-3.2.1.jar

图 2-16　Referenced Libraries 文件夹结构

2.2.2　res 文件夹

本节主要介绍 res 文件夹下的各类资源。

（1）res/anim：包含用 XML 语言描述的应用程序使用的动画效果的配置文件。

（2）res/values：存放字符串、颜色、字体等常量数据。

colors.xml 定义 colordrawable 和颜色的字符串值（color、string、values）。使用 Resource.getDrawable()和 Resources.getColor()分别获得这些资源。

dimens.xml 定义尺寸值（dimension value）。使用 Resources.getDimension()获得这些资源。

strings.xml 定义字符串（string）值。使用 Resources.getString()或者 Resources.getText()获取这些资源。getText()会保留在 UI 字符串上应用的丰富的文本样式。下面通过创建的 Chapter02 工程来演示如何使用 Android 中的字符串资源。Chapter02 工程存放字符串资源的位置在 res 的 values 文件夹下，如图 2-17 所示。

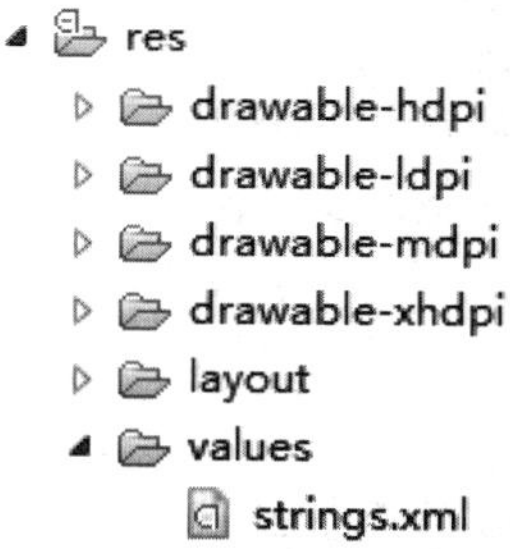

```
▲ 🗁 res
    ▷ 🗁 drawable-hdpi
    ▷ 🗁 drawable-ldpi
    ▷ 🗁 drawable-mdpi
    ▷ 🗁 drawable-xhdpi
    ▷ 🗁 layout
    ▲ 🗁 values
        📄 strings.xml
```

图 2-17 字符串资源文件

Chapter02 中 strings.xml 文件内容如下所示：

```
1.  <?xml version="1.0" encoding="utf-8"?>
2.  <resources>
3.
4.  <string name="content">图书管理系统</string>
5.  <string name="app_name">BookSystem</string>
6.
7.  </resources>
```

当在此创建或修改资源文件时，ADT 都会自动更新 R.java，并产生唯一的标识符来标识，如：

```
1.  public static final class string {
2.      public static final int app_name=0x7f040000;
3.      public static final int menu_settings=0x7f040002;
4.      public static final int title_Activity_main=0x7f040003;
5.  }
```

在程序中使用资源就可以用 R.string 来标识字符串了，并可用 Activity 中的 getText(R.string)直接转成字符串。另外在其他的 XML 文件（如布局文件 Activity_main.xml）中使用字符串可以定义如下："@string/字符串的名称"，这里的名称为 strings.xml 文件中的<string name="字符串的名称">所定义的。

AndroidManifest.xml 中使用字符串的代码如下：

```
1.  <application
2.      android:icon="@drawable/launch"
3.      android:label="@string/app_name"
4.      >
```

```
5.        <Activity
6.            android:name=".MainActivity"
7.            android:label="@string/title_Activity_main" >
8.            <intent-filter>
9.                <action android:name="android.intent.action.MAIN" />
10.
11.               <category android:name="android.intent.category.LAUNCHER"/>
12.           </intent-filter>
13.       </Activity>
14.   </application>
```

以上就是字符串资源的用法。

（3）res/layout：存放一些与 UI 相应的布局文件，都是 XML 文件，如图 2-18 所示。

```
▲ 🗁 layout
      📄 main.xml
```

图 2-18 Layout 文件夹结构

☞**注意**：与其他的 res 文件夹不同，它可以保存任意数量的文件，这些文件保存了要创建资源的描述，而不是资源本身。XML 元素类型控制这些资源应该放在 R 类的什么地方。这个文件夹中的文件可以任意命名，下面给出一些比较典型的文件（文件命名的惯例是将元素类型包含在该名称之中），例如，array.xml 用来定义数组。

图片资源的使用与字符串类似。Chapter02 工程存放图片资源的位置在 res 下的 drawable 文件夹里，如图 2-19 所示。

图 2-19 图片资源

（4）res/drawable：放置应用到的图片资源。其中 drawable-xhdpi 中主要放超高分辨率的图片（有的工程有此文件，有的没有）；drawable-hdpi 中主要放高分辨率的图片；drawable-mdpi 里面主要放中等分辨率的图片；drawable-ldpi 中主要放低分辨率的图片。系统会根据机器的分辨率来分别到这几个文件夹中去找对应的图片，所以在开发程序时为了兼容不同平台、不同屏幕,各文件夹应根据需求存放不同版本的图片。

在 R.java 文件中生成了图片资源的唯一标识,代码如下:

```
1.   public static final class drawable {
2.       public static final int launch=0x7f020000;
3.   }
```

在代码中可以通过 R.drawable.图片名称得到图片资源,在 XML 文件中可通过如下定义活动图片资源:"@drawable/图片名称"。在 AndroidManifest.xml 中使用了图片资源为 Chapter02 工程设定应用图像,设定代码如下:

```
1.   
2.   <application
3.        android:icon="@drawable/launch"
4.        android:label="@string/app_name" >
```

图片资源的用法类似于字符串,都会在 R.java 文件中生成唯一的标识符,在代码和 XML 文件中均可被访问。

运行 Chapter02,BookSystem 为应用,如图 2-20 所示。

图 2-20　BookSystem 应用

2.2.3 AndroidManifest.xml 文件

1. AndroidManifest.xml 的概念

AndroidManifest.xml 是每个 Android 程序中必需的文件。它位于整个项目的根文件夹下，描述了 package 中暴露的组件（activities、services 等）、它们各自的实现类以及各种能被处理的数据和启动位置。除了能声明程序中的 Activities、ContentProviders、Services 和 Intent Receivers 外，还能指定 permissions 和 instrumentation（安全控制和测试）。

2. AndroidManifest.xml 的结构

下面是本章的 AndroidManifest.xml 文件的代码：

```
1.  <?xml version="1.0" encoding="utf-8"?>
2.  <manifest xmlns:android="http://schemas.android.com/apk/res/android"
3.  package="my.system"
4.  android:versionCode="1"
5.  android:versionName="1.0" >
6.  <uses-sdk android:minSdkVersion="15" />
7.  <application
8.  android:icon="@drawable/ic_launcher"
9.  android:label="@string/app_name" >
10. <Activity
11. android:name=".Chapter02Activity"
12. android:label="@string/app_name" >
13. <intent-filter>
14. <action android:name="android.intent.action.MAIN" />
15. <category android:name="android.intent.category.LAUNCHER" />
16. </intent-filter>
17. </Activity>
18. </application>
19. </manifest>
```

下面对上面的部分代码进行解释说明。

（1）Manifest：是根节点，描述了 package 中所有的内容。

（2）xmlns:android：定义 Android 命名空间，一般为 http://schemas.android.aom/apk/res/android，这样使得 Android 中各种标准属性能在文件中使用，提供了大部分元素中的数据。

(3) package：指定本应用内 Java 主程序包的包名，它也是一个应用进程的默认名称。

(4) android:versionCode：是给设备程序识别版本（升级）用的，必须是一个 interger 值，代表 app 更新过多少次，比如第一版一般为 1，之后若要更新版本就设置为 2，以此类推。

(5) android：versionName：这个名称是给用户看的，可以将你的 APP 版本号设置为 1.1 版，后续更新版本设置为 1.2、2.0 版本等。

(6) <uses-sdk />：

1. <uses-sdk android:minSdkVersion="integer"
2. android:targetSdkVersion="integer"
3. android:maxSdkVersion="integer"/>

描述应用所需的 api level，就是版本，目前是 android 2.2=8、android2.1=7、android1.6=4、android1.5=3。在此属性中可以指定支持的最小版本、目标版本以及最大版本。

(7) android:icon：这个很简单，就是声明整个 APP 的图标，图片一般都放在 drawable 文件夹下。

(8) android:label：这个属性用于给内容提供器定义一个用户可读的标签。如果没有设置这个属性，那么它会使用<application>元素的 label 属性值来代替。

(9) <Application>：一个 AndroidManifest.xml 中必须含有一个 Application 标签，这个标签声明了每一个应用程序的组件及其属性（如 icon、label、permission 等）。

(10) <intent-filter>：这个元素用于指定 Activity、Service 或 Broadcast Receiver 能够响应的 Intent 对象的类型。Intent 过滤器声明了它的父组件的能力 Activity 或 Service 所能做的事情和 Broadcast Receiver 所能够处理的广播类型。它会打开组件来接收它所声明类型的 Intent 对象，过滤掉那些对组件没有意义的 Intent 对象请求。过滤器的大多数内容是通过<action>、<category>和<data>子元素来描述的。

(11) Action：组件支持的 Intentaction。

(12) Category：组件支持的 IntentCategory。

2.3 最简单的图书管理系统

本节介绍一个单机版的、无网络的、无数据库的、只有一个界面的图书管理系统。首先展示 Chapter02 程序在模拟器上的运行结果如图 2-21 所示。在这个程序中运用了本章学习的 4 个知识点：字符串、图片、布局文件和 Activity 的生命周期。

新建一个程序，以 Chapter02 命名，以 MainActivity.java 为类名，得到如图 2-22 所示的程序文件夹。

图 2-21 运行结果　　图 2-22 示例代码文件结构

下面分析代码。

首先是 ui.cqupt 包下的 MainActivity.java，代码如下所示：

```
1.  package ui.cqupt;
2.
3.  import android.os.Bundle;
4.  import android.app.Activity;
5.
6.  public class MainActivity extends Activity {
7.
8.      public void onCreate(Bundle savedInstanceState) {
9.          super.onCreate(savedInstanceState);
10.         setContentView(R.layout.main);
11.     }
12.
13. }
14.
```

在 MainActivity 首先覆写了 onCreate()，通过 setContentView(R.layout.main)设定了当

前 Activity 显示的布局为 main.xml。

main.xml 为布局文件，代码如下所示：

```
1.  <LinearLayout xmlns:android="http://schemas.android.com/apk/res/android"
2.      xmlns:tools="http://schemas.android.com/tools"
3.      android:layout_width="match_parent"
4.      android:layout_height="match_parent" >
5.  
6.      <TextView
7.          android:textSize="15pt"
8.          android:layout_width="fill_parent"
9.          android:layout_height="wrap_content"
10.         android:text="图书管理系统" />
11. 
12. </LinearLayout>
```

main.xml 布局是线性布局（Linear Layout），图 2-21 中的"图书管理系统"是在一个 TextView 组件中显示，如代码的第 6~10 行，在 TextView 中定义了此 TextView 的属性，第 7 行设定了 TextView 中字体的大小为 15px，第 8、9 行设置了 TextView 的宽度和高度，第 10 行是设置 TextView 中显示的内容。

图 2-20 中的 BookSystem 和图标是本应用的名称和标识，运用了字符串和图片的知识，在 AndroidManifest.xml 设定的，AndroidManifest.xml 代码如下所示：

```
1.  <manifest xmlns:android="http://schemas.android.com/apk/res/android"
2.      package="ui.cqupt"
3.      android:versionCode="1"
4.      android:versionName="1.0" >
5.  
6.      <uses-sdk
7.          android:minSdkVersion="8"
8.          android:targetSdkVersion="15" />
9.  
10.     <application
11.         android:icon="@drawable/launch"
12.         android:label="@string/app_name" >
13. 
14.         <Activity
15.             android:name=".MainActivity"
16.             android:label="@string/title_Activity_main" >
17.             <intent-filter>
18.                 <action android:name="android.intent.action.MAIN" />
19. 
20.                 <category android:name="android.intent.category.LAUNCHER"/>
21.             </intent-filter>
```

```
22.        </Activity>
23.    </application>
24.
25. </manifest>
```

代码的第 11 行设定了图书管理系统的图标，在 R.class 文件中生成的代码如下所示：

```
1. public static final class drawable {
2.       public static final int launch=0x7f020000;
3.   }
```

代码的第 12、16 行通过字符串设定了应用的标签和 MainActivity 的标签。string.xml 代码如下所示：

```
1. <resources>
2.
3.    <string name="app_name">Chapter02</string>
4.    <string name="content">图书管理系统</string>
5.    <string name="title_Activity_main">BookSystem</string>
6.
7. </resources>
```

strings.xml 在 R.class 中生成的代码如下所示：

```
1.   public static final class string {
2.       public static final int app_name=0x7f040000;
3.       public static final int content=0x7f040001;
4.       public static final int title_Activity_main=0x7f040003;
5.   }
```

这就是本章的实例——最简单的图书管理系统。希望大家通过本章的学习对 Activity 和工程的整体结构有深入的了解。

第 3 章 用户界面

3.1 用户界面基础知识

对于一个应用程序来说,首先呈献给用户的肯定是用户界面(User Interface,UI),所以用户界面对一个应用程序来说是非常重要的。

程序的用户界面是指用户看到的并与之交互的一切,Android 提供了一个强大的模式来定义用户界面,这个模式基于基础的布局类:视图(View)和视图组(ViewGroup)。Android 提供了多种预先生成的视图和视图组的子类,用于构建你自己的用户界面。其实 ViewGroup 也是继承了 View 的,所以 Android 预先生成的这些组件都是 View 类的子类。

用户界面的设计可分为两种方式:一种设计方式是单一的采用代码的方式进行用户界面的设计,例如编写一个 Java 的 Swing 应用,你需要用 Java 代码去创建和操纵界面 JFrame 和 JButton 等对象;另一种设计方式是像设计网页时采用类似 XML 的 HTML 标记语言去描述你想看到的页面效果。Android 采取这两种方法均可,让你可以选择任意一种方式进行界面设计,既可以使用 Java 代码也可以采用 XML 声明你想要的界面效果。如果查看 Android 的用户界面 API 文档,你会同时看到 Java 的方法和对应的 XML 属性(如图 3-1 所示)。

XML Attributes		
Attribute Name	Related Method	Description
android:cacheColorHint		Indicates that this list will always be drawn on top of solid, single-color opaque background.
android:choiceMode		Defines the choice behavior for the view.
android:drawSelectorOnTop	setDrawSelectorOnTop(boolean)	When set to true, the selector will be drawn over the selected item.
android:fastScrollEnabled		Enables the fast scroll thumb that can be dragged to quickly scroll through the list.
android:listSelector	setSelector(int)	Drawable used to indicate the currently selected item in the list.
android:scrollingCache		When set to true, the list uses a drawing cache during scrolling.
android:smoothScrollbar	setSmoothScrollbarEnabled(boolean)	When set to true, the list will use a more refined calculation method based on the pixels height of the items visible on screen.
android:stackFromBottom		Used by ListView and GridView to stack their content from the bottom.
android:textFilterEnabled		When set to true, the list will filter results as the user types.
android:transcriptMode		Sets the transcript mode for the list.

图 3-1 Java 方法与对应的 XML 属性表

那到底使用哪一种好点呢？其实两者都可以，但 Google 建议尽量使用 XML，因为相比 Java 代码而言，XML 语句更简短易懂，在以后的版本中不易被改变。本书的所有程序也都基本采用 XML 进行界面设计。

3.2　界面基本组件

在本章开头提到 Android 提供了很多预先生成的组件，下面大致介绍一下这些组件。我们知道，其实每个组件都是视图类（View）的子类，所以每个组件都是一个视图，继承了多种 View 的属性。

3.2.1　界面基本属性

在界面设计时，我们需要对不同的组件设置不同的属性，但上面讲到其实每个组件都是 View 的子类，那么它们肯定会有一些相同的基本属性。下面就列举一些常用的属性。

- android:layout_width：设置组件的宽度。
- android:layout_height：设置组件高度。
- android:background：设置组件的背景，在代码中可以使用 setBackgroundResource() 来设置该属性。
- android:onClick：为组件添加点击事件响应函数，有关事件响应的知识将在 3.5 节中具体讲解。
- android:id：设置组件的 id，同样在代码中使用 setId()方法可达到同样的效果。

界面的基本属性不止这些，而且不同组件拥有自己的特殊属性，比如 LinearLayout 的 android:orientation 属性设置布局的方位是水平还是垂直，这些都将在接下来的组件和布局的讲解中涉及。

3.2.2　TextView

TextView 是标准的只读标签。它支持多行显示，支持字符串格式化和自动换行。对于 TextView 我们最关心的就是怎样设置显示的文本，怎样设置字体的大小、颜色和样式，TextView 提供的大量属性可帮我们轻松地完成这些，图 3-2 就是利用这些属性完成的一个 TextView。

图 3-2 的 XML 布局如下：

```
1.   <?xml version="1.0" encoding="utf-8"?>
2.   <LinearLayout xmlns:android="http://schemas.android.com/apk/res/android"
3.       android:orientation="vertical"
4.       android:layout_width="fill_parent"
5.       android:layout_height="fill_parent"
6.       >
7.   <TextView
```

```
8.        android:layout_width="fill_parent"
9.        android:layout_height="wrap_content"
10.       android:textColor="#fff000"
11.       android:textSize="20dp"
12.       android:textStyle="bold"
13.       android:text="我是文本框"
14.     />
15. </LinearLayout>
```

Activity 的代码如下：

```
1. public class TextView extends Activity{
2.      public void onCreate(Bundle savedInstanceState) {
3.        super.onCreate(savedInstanceState);
4.        setContentView(R.layout.textview);
5.     }}
```

这里增加了 3 个属性的设置，分别是 android:textColor="#fff000" 设置字体为黄色，android:textSize="20dp" 设置字体为 20dp，android:textStyle="bold" 设置字体加粗。

本部分代码见本书配套资源中的工程 Chapter3.2.1。

我们都见过 HTMl 中只要加个 <a/> 标记就可以将一段文字变成超链接的形式，可以单击访问链接地址。TextView 中也有超链接形式。TextView 提供了 android:autoLink 属性，只要把它设置成 web，该 TextView 中的网址形式的文件就会自动变成超链接的形式，如图 3-3 所示。

图 3-2　TextView

图 3-3　网址超链接

图 3-3 的 XML 布局如下：

```
1.  <?xml version="1.0" encoding="utf-8"?>
2.  <LinearLayout xmlns:android="http://schemas.android.com/apk/res/android"
3.      android:layout_width="fill_parent"
4.      android:layout_height="fill_parent"
5.      android:orientation="vertical" >
6.      <TextView
7.          android:id="@+id/text_view"
8.          android:layout_width="fill_parent"
9.          android:layout_height="wrap_content"
10.         android:autoLink="web"
11.         android:text="重庆邮电大学网址:www.cqupt.edu.cn " />
12. </LinearLayout>
```

Activity 的代码如下：

```
1.  public class TextView extends Activity{
2.      public void onCreate(Bundle savedInstanceState) {
3.       super.onCreate(savedInstanceState);
4.       setContentView(R.layout.textview);
5.  }}
```

这里是将网址以超链接显示，如果要将电话号码显示为超链接，那么只需将 android:autoLink 属性设置为 phone，E-mail 也是同理。那能不能将网址、电话、E-mail 都设为超链接形式呢？当然可以，只需将 android:autoLink 属性设为 all，这样里面的网址、电话和 E-mail 就都显示为超链接。

当然我们经常会在代码中修改这些属性，那时只需要调用这些属性对应的方法即可，如 android:textColor 对应的 setTextColor(int) 方法、android:autoLink 对应的 setAutoLinkMask(int)方法等，这里就不一一列举了，读者可查看 TextView 的 API 了解其他属性设置。

本部分代码见本书配套资源中的工程 Chapter3.2.2。

3.2.3 EditText

EditText 是 TextView 的子类，它与 TextView 一样具有支持多行显示、字符串格式化和自动换行的功能，它的使用和 TextView 并无太大区别。但在实际编程中经常要求编辑框输入一些特定的内容，例如 0～9 的数字、E-mail 等，如图 3-4 所示。接下来要实现的就是这个实例。

图 3-4　不同输入类型的 EditText

图 3-4 的 XML 布局如下：

```
1.   <?xml version="1.0" encoding="utf-8"?>
2.   <LinearLayout xmlns:android="http://schemas.android.com/apk/res/android"
3.       android:layout_width="fill_parent"
4.       android:layout_height="fill_parent"
5.       android:orientation="vertical" />
6.     <TextView
7.         android:layout_width="fill_parent"
8.         android:layout_height="wrap_content"
9.         android:text="使用 android:inputType 属性，输入 E-mail" />
10.    <EditText
11.        android:layout_width="fill_parent"
12.        android:layout_height="wrap_content"
13.        android:inputType="textE-mailAddress" />
14.    <TextView
15.        android:layout_width="fill_parent"
16.        android:layout_height="wrap_content"
17.        android:text="使用 android:digits 属性，输入 26 个小写字母" />
18.    <EditText
19.        android:layout_width="fill_parent"
20.        android:layout_height="wrap_content"
21.        android:digits="abcdzfghijklmnopqrstuvwxyz" />
22.    <TextView
23.        android:layout_width="fill_parent"
```

```
24.         android:layout_height="wrap_content"
25.         android:text="使用 android:numeric 属性,输入 0~9 数字" />
26.     <EditText
27.         android:layout_width="fill_parent"
28.         android:layout_height="wrap_content"
29.         android:numeric="integer" />
30. </LinearLayout>
```

Activity 的代码如下:

```
1. public class TextView extends Activity{
2.      public void onCreate(Bundle savedInstanceState) {
3.       super.onCreate(savedInstanceState);
4.       setContentView(R.layout.textview);
5.     }}
```

上面这段代码中使用了 EditText 的 3 个属性指定了 3 种不同的输入字符,分别是:

- 将 android:inputType 的属性值设置为 textEmailAddress,指定输入为 E-mail。但要注意的是用于输入 E-mail 的 EditText 逐渐不会限制输入非 E-mail 字符,只是在虚拟键盘上多了一个 "@" 键。
- 将 android:digits 属性值设为 26 个小写英文字母,将输入内容限制在 26 个小写字母内。
- 将 android:numeric 属性值设为 integer,设置其内容为整数。

关于 3 个标签的其他属性可以查阅官方的参考文档。本节代码见本书配套资源中的工程 Chapter3.2.3。

3.2.4 Button

Button 是最常见的界面组件,在 Android 中也不例外。Android 除了提供一些典型的按钮外,还提供了一些额外的按钮。本节将讲解 3 种不同的按钮:普通按钮(Button)、图片按钮(ImageButton)、开关按钮(ToggleButton)。图 3-5 展示了这 3 种按钮的效果图。最上面的是普通按钮,中间的是图片按钮,最下面的是开关按钮,当前处于关闭状态。

1. 普通按钮(Button)

关于普通按钮,除了掌握单击事件(click event)外没有太多的知识点。

首先使用下面的代码定义一个普通的 Button。

```
1. <Button
2.     android:id="@+id/button"
3.     android:layout_width="wrap_content"
4.     android:layout_height="wrap_content"
5.     android:text="普通按钮"/>
```

图 3-5 3 种按钮

接下来就是为该 Button 添加按钮单击事件，只需要调用 Button 的 setOnClickListener 函数，OnClickListener 接口的子类为参数，在 onClick 中做出事件的响应。有关监听事件的讲解将在 3.5 节进行。事件监听代码如下：

```
1.   Button btn = (Button) this.findViewById(R.id.button);
2.          btn.setOnClickListener(new OnClickListener() {
3.              public void onClick(View v) {
4.                  Toast.makeText(MainActivity.this, "button",
                            Toast.LENGTH_SHORT).show();
5.              }
6.          });
```

2. 图片按钮（ImageButton）

Android 在 android.widget 包定义了一个 ImageButton。ImageButton 的用法除了要为按钮添加图片资源外其他用法与普通按钮的用法没有太大差别，ImageButton 的定义如下：

```
1.   <ImageButton
2.       android:id="@+id/imageButton1"
3.       android:layout_width="wrap_content"
4.       android:layout_height="wrap_content"
5.       android:src="@drawable/ic_launcher" />
```

与 Button 的定义比较，ImageButton 的定义只是多了个 android:src 属性为 ImageButton 添加图片资源，此外还可以在 Java 代码中调用 setImageURI (Uri uri)方法添加图片资源。

关于 ImageButton 的其他属性读者可以参考官方文档。

3．开关按钮（ToggleButton）

ToggleButton 是 Android 定义的一种特殊按钮，它有两种状态：开（On）、关（Off）。ToggleButton 的默认状态是在开启状态时显示一条绿色的进度条，关闭时显示一条灰色的滚动条。下面的代码展示了一个 ToggleButton 的定义。

```
1.  <ToggleButton
2.      android:id="@+id/toggleButton"
3.      android:layout_width="wrap_content"
4.      android:layout_height="wrap_content"
5.      android:text="开关按钮"
6.      android:textOff="Stop"
7.      android:textOn="Run" />
```

这里的 android:textOff 和 android:textOn 属性设置了默认状态下开启和关闭状态时显示的文字。ToggleButton 在其他方面的应用与 Button 也没有太大的区别，其双状态功能的特性在开关等功能时较有用处。

本节代码见本书配套资源中的工程 Chapter3.2.4。

3.2.5 复选框（CheckBox）

复选框（CheckBox）在所有小部件工具包都有涉及。HTML、图形用户界面和 JSF 都支持 CheckBox 这个概念。同开关按钮一样，CheckBox 是一种拥有两种状态的按钮，它允许用户在两种状态间自由切换。下面就来在 Android 中创建一个 CheckBox，如图 3-6 所示。

图 3-6　CheckBox

```
1.  <LinearLayout xmlns:android="http://schemas.android.com/apk/res/android"
2.      android:layout_width="fill_parent"
3.      android:layout_height="fill_parent"
4.      android:orientation="vertical" >
5.      <CheckBox
6.          android:layout_width="wrap_content"
7.          android:layout_height="wrap_content"
8.          android:text="语文" />
9.      <CheckBox
10.         android:layout_width="wrap_content"
11.         android:layout_height="wrap_content"
12.         android:text="数学" />
13.     <CheckBox
14.         android:layout_width="wrap_content"
15.         android:layout_height="wrap_content"
16.         android:text="外语" />
17. </LinearLayout>
```

在编程中可以通过 setChecked()或 toggle()来管理一个 CheckBox，通过 isChecked()来获得一个复选框的状态。

如果当一个复选框被选中或取消选中时需要实现特定的逻辑，可以通过 setOnCheckedChangeListener()为复选框注册一个实现了 OnCheckedChangeListener 的类，并实现 OnCheckedChangeListener 的 onCheckedChanged()方法，这个方法会在复选框的状态改变时调用。有关 CheckBox 的知识读者可参考官方文档 android.widget.CheckBox。

本节代码见本书配套资源中的工程 **Chapter3.2.5**。

3.2.6 单选按钮

单选按钮（RadioButton）控件是任何 UI 工具包的组成部分，它为用户提供了单击选中并要求必须选中一个项目的功能。为了完成这个单选功能，每个单选按钮须归属于一个组，每个组在每段时间只能有一个选项被选中，如图 3-7 所示。

RadioButton 的定义如下：

```
1.  <LinearLayout xmlns:android=http://schemas.android.com/apk/res/android
2.
3.      android:layout_width="fill_parent"
4.      android:layout_height="fill_parent"
5.      android:orientation="vertical" >
6.      <RadioGroup
7.          android:id="@+id/rBtnGrp"
8.          android:layout_width="wrap_content"
9.          android:layout_height="wrap_content"
10.         android:orientation="vertical" >
11.         <RadioButton
```

```
12.         android:id="@+id/RBtn1"
13.         android:layout_width="wrap_content"
14.         android:layout_height="wrap_content"
15.         android:checked="true"
16.         android:text="语文" />
17.     <RadioButton
18.         android:id="@+id/RBtn2"
19.         android:layout_width="wrap_content"
20.         android:layout_height="wrap_content"
21.         android:text="数学" />
22.     <RadioButton
23.         android:id="@+id/RBtn3"
24.         android:layout_width="wrap_content"
25.         android:layout_height="wrap_content"
26.         android:text="英语" />
27.     </RadioGroup>
28. </LinearLayout>
```

图 3-7　RadioButton

3 个单选按钮都放在一个单选组里面，并将第一个单选按钮的 android:checked 属性设为 true，这样第一个按钮就默认被选中，android:checked 在默认情况下是 false。RadioButton 的管理和事件监听处理和与 CheckBox 是一致的，此处不再赘述。

本节代码见本书配套资源中的工程 Chapter3.2.6。

3.2.7 Listview

最后介绍一个在 Android 开发中非常重要的一个组件——Listview。ListView 是一个 ViewGroup，用于创建一个滚动的项目清单，通过使用 ListAdapter，列表中的项目会自动插入到列表中。本节将创建一个简单的滚动图书列表，当单击了一项后就会弹出一个 Toast（即一闪而过的简单提示框。如图 3-8 屏幕下方"美术"处的提示框）提示单击的图书。具体效果如图 3-8 所示。

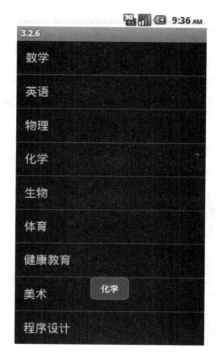

图 3-8　图书列表

具体创建步骤如下。

（1）创建工程 BookListView。

（2）创建一个 XML 布局文件 list_item.xml，该文件定义了每个将被放置在 ListView 中的项目的布局。list_item.xml 的内容如下：

```
1.  <?xml version="1.0" encoding="utf-8"?>
2.  <TextView xmlns:android=http://schemas.android.com/apk/res/android
3.      android:layout_width="fill_parent"
4.      android:layout_height="fill_parent"
5.      android:padding="15dp"
6.      android:textSize="20px" >
7.  </TextView>
```

（3）打开 BookListViewActivity.Java，让 BookListViewActivity 类继承 ListActivity。BookListViewActivity 的代码如下：

```
1.  public class BookListViewActivity extends ListActivity {
2.      public void onCreate(Bundle savedInstanceState) {
3.          super.onCreate(savedInstanceState);
4.          setListAdapter(new ArrayAdapter<String>(this, R.layout.list_item,
5.              BOOKS));
6.          ListView lv = getListView();
7.          lv.setTextFilterEnabled(true);
8.          lv.setOnItemClickListener(new OnItemClickListener() {
9.              public void onItemClick(AdapterView<?> parent, View view,
10.                 int position, long id) {
11.                 Toast.makeText(getApplicationContext(),
12.                     ((TextView) view).getText(),Toast.LENGTH_
                        SHORT).show();
13.             }
14.         });
15.     }
16.     private static final String[] BOOKS = new String[] {"数学",
            "英语","物理","化学","生物","体育","健康教育","美术","程序设计",
            "Android程序教程"};
17. }
```

这样简单的图书列表就完成了，但只是简单地使用 Listview 显示了一行文字，在实际开发中每个项目的布局要比这复杂得多，这就需要修改 list_item.xml 的布局并使用合适的 ListAdapter。

本节代码见本书配套资源中的工程 Chapter3.2.7。

3.3 布　　局

Android 拥有许多的组件，而 TextView、EditText 和 Button 就是其中比较常用的几个组件。那么 Android 又是怎样将组件简洁而又美观地分布在界面上的呢？这里就用到了 Android 的布局管理器。这些独立的组件通过 Android 布局组合到一起，就可以给用户提供复杂而有序的界面。

3.3.1 FrameLayout（帧布局）

FrameLayout 是从屏幕的左上角的（0,0）坐标开始布局，多个组件层叠排列，第一个添加的组件放到最底层，最后添加到框架中的视图显示在最上面。上一层的会覆盖下一层的控件。

FrameLayout 是最简单的一个布局对象。它被定制为屏幕上的一个空白备用区域，之后你可以在其中填充一个单一对象。例如，一张要发布的图片。所有的子元素将会固定在屏幕的左上角；你不能为 FrameLayout 中的一个子元素指定一个位置。后一个子元素将会直接在前一个子元素之上进行覆盖填充，把它们部分或全部挡住（除非后一个子元素是透明的）。图 3-9 就是一个 FrameLayout 的布局效果。

图 3-9　FrameLayout（帧布局）

图 3-9 对应的代码如下：

```
1.  <?xml version="1.0" encoding="utf-8"?>
2.  <FrameLayout xmlns:android="http://schemas.android.com/apk/res/android"
3.      android:layout_width="fill_parent"
4.      android:layout_height="fill_parent" >
5.      <TextView
6.          android:layout_width="300dp"
7.          android:layout_height="200dp"
8.          android:background="#F11EEE" />
9.      <TextView
10.         android:layout_width="260dp"
11.         android:layout_height="160dp"
12.         android:background="#FFFFFF" />
13.     <TextView
14.         android:layout_width="220dp"
15.         android:layout_height="120dp"
16.         android:background="#000FFF" />
17. </FrameLayout>
```

完整代码见本书配套资源中的工程 Chapter3.3.1。

3.3.2 LinearLayout（线性布局）

LinearLayout 以为它设置的垂直或水平的属性值来排列所有的子元素。所有的子元素都被堆放在其他元素之后，因此一个垂直列表的每一行只会有一个元素，而不管它们有多宽，一个水平列表将会只有一个行高（高度为最高子元素的高度加上边框高度）。LinearLayout 保持子元素之间的间隔并互相对齐（相对一个元素的右对齐、中间对齐或者左对齐）。

LinearLayout 还支持为单独的子元素指定 weight。好处就是允许子元素可以填充屏幕上的剩余空间。这也避免了在一个大屏幕中，一串小对象挤成一堆的情况，而是允许它们放大填充空白。子元素指定一个 weight 值，剩余的空间就会按这些子元素指定的 weight 比例分配。默认的 weight 值为 0。例如，如果有 3 个文本框，其中两个指定了 weight 值为 1，那么，这两个文本框将等比例地放大，并填满剩余的空间，而第三个文本框不会放大。

线性布局是 Android 开发中最常见的一种布局方式，它是按照垂直或者水平方向来布局的，通过 android:orientation 属性可以设置线性布局的方向。属性值有垂直（vertical）和水平（horizontal）两种。

常用的属性如下：
- android:orientation——可以设置布局的方向。
- android:gravity——用来控制组件的对齐方式。
- layout_weight——控制各个组件在布局中的相对大小。

如图 3-10 就是一个 LinearLayout 的布局效果。

图 3-10　LinearLayout（线性布局）

图 3-10 对应的代码如下：

```
1.  <?xml version="1.0" encoding="utf-8"?>
2.  <LinearLayout xmlns:android="http://schemas.android.com/apk/res/android"
3.      android:orientation="vertical"
4.      android:layout_width="fill_parent"
5.      android:layout_height="fill_parent" >
6.      <LinearLayout
7.          android:layout_width="fill_parent"
8.          android:layout_height="wrap_content"
9.          android:orientation="vertical"
10.         >
11.     <TextView
12.         android:layout_width="fill_parent"
13.         android:layout_height="wrap_content"
14.         android:text="请输入用户名："
15.         android:textSize="10pt" />
16.     <EditText
17.         android:layout_width="fill_parent"
18.         android:layout_height="wrap_content" />
19.     <TextView
20.         android:layout_width="fill_parent"
21.         android:layout_height="wrap_content"
22.         android:text="请输入密码："
23.         android:textSize="10pt" />
24.     <EditText
25.         android:layout_width="fill_parent"
26.         android:layout_height="wrap_content" />
27.     </LinearLayout>
28.     <LinearLayout
29.         android:layout_width="fill_parent"
30.         android:layout_height="wrap_content"
31.         android:orientation="horizontal"
32.         android:gravity="right"
33.         >
34.     <!-- android:gravity="right"表示 Button 组件向右对齐 -->
35.         <Button
36.             android:layout_height="wrap_content"
37.             android:layout_width="wrap_content"
38.             android:text="确定"
39.             />
40.         <Button
41.             android:layout_height="wrap_content"
42.             android:layout_width="wrap_content"
```

```
43.            android:text="取消"
44.        />
45.    </LinearLayout>
46. </LinearLayout>
```

完整代码见本书配套资源中的工程 Chapter3.3.2。

3.3.3 RelativeLayout（相对布局）

RelativeLayout 允许子元素指定它们相对于其他元素或父元素的位置（通过 ID 指定）。因此，可以以右对齐、或上下、或置于屏幕中央的形式来排列两个元素。元素按顺序排列，因此如果第一个元素在屏幕的中央，那么相对于这个元素的其他元素将以屏幕中央的相对位置来排列。如果使用 XML 来指定这个 layout，那么在定义它之前，必须先定义被关联的元素。

图 3-11 就是一个 RelativeLayout 的效果。

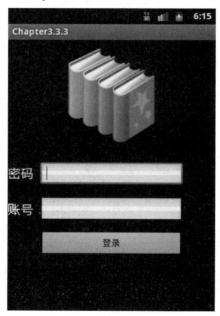

图 3-11　**RelativeLayout**（相对布局）

图 3-11 对应的代码如下：

```
1.  <?xml version="1.0" encoding="utf-8"?>
2.  <RelativeLayout xmlns:android="http://schemas.android.com/apk/res/android"
3.      android:layout_width="fill_parent"
4.      android:layout_height="fill_parent"
5.      android:orientation="vertical" >
6.      <EditText
7.          android:id="@+id/user"
8.          android:layout_width="220dip"
```

```
9.          android:layout_height="40dip"
10.         android:layout_centerHorizontal="true"
11.         android:layout_centerVertical="true"
12.         android:inputType="text"
13.         android:textColor="#000000" />
14.     <EditText
15.         android:id="@+id/password"
16.         android:layout_width="220dip"
17.         android:layout_height="40dip"
18.         android:layout_alignLeft="@+id/user"
19.         android:layout_below="@+id/user"
20.         android:layout_marginTop="16dp"
21.         android:inputType="textPassword"
22.         android:textColor="#000000" />
23.     <Button
24.         android:id="@+id/button_login"
25.         android:layout_width="220dip"
26.         android:layout_height="40dip"
27.         android:layout_alignLeft="@+id/password"
28.         android:layout_below="@+id/password"
29.         android:layout_marginTop="14dp"
30.         android:text="登录"
31.         android:textColor="#000000" />
32.     <TextView
33.         android:id="@+id/textView_user"
34.         android:layout_width="fill_parent"
35.         android:layout_height="wrap_content"
36.         android:layout_alignBaseline="@+id/password"
37.         android:layout_alignBottom="@+id/password"
38.         android:layout_alignParentLeft="true"
39.         android:text="账号"
40.         android:textColor="#FFFFFF"
41.         android:textSize="20dp" />
42.     <TextView
43.         android:id="@+id/textView_password"
44.         android:layout_width="fill_parent"
45.         android:layout_height="wrap_content"
46.         android:layout_alignBaseline="@+id/user"
47.         android:layout_alignBottom="@+id/user"
48.         android:layout_alignParentLeft="true"
49.         android:text="密码"
50.         android:textColor="#FFFFFF"
51.         android:textSize="20dp" />
52.     <ImageView
53.         android:id="@+id/imageView_login"
```

```
54.         android:layout_width="wrap_content"
55.         android:layout_height="wrap_content"
56.         android:layout_above="@+id/user"
57.         android:layout_centerHorizontal="true"
58.         android:layout_marginBottom="22dp"
59.         android:src="@drawable/yu" />
60. </RelativeLayout>
```

完整代码见本书配套资源中的工程 Chapter3.3.3。

3.3.4 TableLayout（表格布局）

TableLayout 将子元素的位置分配到行或列中。一个 TableLayout 由许多的 TableRow 组成，每个 TableRow 都会定义一个 row（事实上，你可以定义其他的子对象）。TableLayout 容器不会显示 rowcloumns 或 cell 的边框线。每个 row 拥有 0 个或多个的 cell；每个 cell 拥有一个 View 对象。表格由列和行组成许多单元格。表格允许单元格为空。单元格不能跨列，这与 HTML 中的不一样。

如图 3-12 所示为 TableLayout 布局效果。

图 3-12 TableLayout（表格布局）

图 3-12 对应的代码如下：

```
1.  <?xml version="1.0" encoding="utf-8"?>
2.  <TableLayout xmlns:android="http://schemas.android.com/apk/res/android"
3.      android:layout_width="fill_parent"
4.      android:layout_height="fill_parent" >
5.      <TableRow>
6.          <Button
```

```
7.            android:text="按钮 1 " />
8.        <Button
9.            android:text="按钮 2" />
10.       <Button
11.           android:text="按钮 3" />
12.    </TableRow>
13.    <TableRow>
14.        <Button
15.            android:layout_span="2"
16.            android:text="按钮 4"/>
17.        <Button
18.            android:text="按钮 5" />
19.    </TableRow>
20. </TableLayout>
```

完整代码见本书配套资源中的工程 Chapter3.3.4。

3.3.5 AbsoluteLayout（绝对布局）

AbsoluteLayout 可以让子元素指定准确的 x/y 坐标值，并显示在屏幕上。(0, 0)为左上角，当向下或向右移动时，坐标值将变大。AbsoluteLayout 没有页边框，允许元素之间互相重叠（尽管不推荐）。手机应用需要适应不同的屏幕大小，而这种布局模型不能自适应屏幕尺寸大小，所以通常不推荐使用 AbsoluteLayout，除非你有正当理由要使用它，因为它使界面代码太过刚性，以至于在不同的设备上可能不能很好地工作。

图 3-13 为 AbsoluteLayout 布局效果。

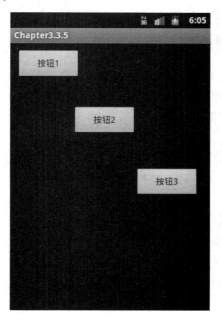

图 3-13　AbsoluteLayout（绝对布局）

图 3-13 对应的代码如下：

```
1.   <?xml version="1.0" encoding="utf-8"?>
2.    <AbsoluteLayout xmlns:android="http://schemas.android.com/apk/res/android"
3.        android:layout_width="fill_parent"
4.        android:layout_height="fill_parent"
5.        android:orientation="vertical" >
6.        <Button
7.            android:layout_width="102dp"
8.            android:layout_height="wrap_content"
9.            android:layout_x="10dp"
10.           android:layout_y="10dp"
11.           android:text="按钮 1" />
12.       <Button
13.           android:layout_width="102dp"
14.           android:layout_height="wrap_content"
15.           android:layout_x="100dp"
16.           android:layout_y="100dp"
17.           android:text="按钮 2" />
18.       <Button
19.           android:layout_width="102dp"
20.           android:layout_height="wrap_content"
21.           android:layout_x="200dp"
22.           android:layout_y="200dp"
23.           android:text="按钮 3" />
24.   </AbsoluteLayout>
```

完整代码见本书配套资源中的工程 Chapter3.3.5。

3.3.6 多种布局混合使用

在 Android 开发中，如果发现一个界面布局文件只使用一种布局不能达到理想的效果，则可混合使用两种或两种以上的布局。图 3-14 就是 Android 的混合布局。

图 3-14 对应的代码如下：

```
1.   <?xml version="1.0" encoding="utf-8"?>
2.   <LinearLayout xmlns:android="http://schemas.android.com/apk/res/android"
3.       android:layout_width="fill_parent"
4.       android:layout_height="fill_parent"
5.       android:orientation="vertical" >
6.       <Button
7.           android:layout_width="102dp"
8.           android:layout_height="wrap_content"
9.           android:text="按钮 1" />
10.      <Button
11.          android:layout_width="102dp"
12.          android:layout_height="wrap_content"
```

```
13.        android:text="按钮 2" />
14.    <Button
15.        android:layout_width="102dp"
16.        android:layout_height="wrap_content"
17.        android:text="按钮 3" />
18.    <RelativeLayout
19.        android:layout_width="fill_parent"
20.        android:layout_height="fill_parent"
21.        android:background="#FFFFFF"
22.        android:orientation="horizontal" >
23.        <ImageView
24.            android:id="@+id/imageView_item_book"
25.            android:layout_width="200dip"
26.            android:layout_height="160dip"
27.            android:src="@drawable/ic_book" />
28.        <TextView
29.            android:id="@+id/information_item"
30.            android:layout_width="220dip"
31.            android:layout_height="wrap_content"
32.            android:layout_alignParentRight="true"
33.            android:layout_below="@+id/imageView_item_book"
34.            android:text="文本框"
35.            android:textColor="#000000"
36.            android:textSize="50dp" />
37.    </RelativeLayout>
38. </LinearLayout>
```

完整代码见本书配套资源中的工程 Chapter3.3.6。

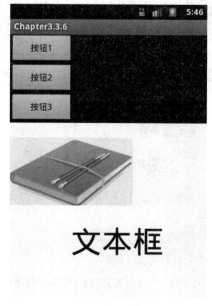

图 3-14 混合布局

3.4 菜　　单

Android 提供了一些简单的方法来为应用添加 Menu 菜单。大致分为 3 种类型：选项菜单、上下文菜单和子菜单。

3.4.1 选项菜单

选项菜单（Options Menu）是最常规的菜单，它是通过 Menu 按钮调用菜单。

如图 3-15 所示，按下手机上的 Menu 按钮（）就会出现图中的菜单。

图 3-15　选项菜单

选项菜单最多只有 6 个，超过 6 个时第六个就会自动显示"更多"选项来展开显示。

具体用法如下：

- 通过 onCreateOptionsMenu()方法只会调用一次，即第一次显示的时候会调用。
- 如果需要更新菜单项，可以在 onPrepareOptionsMenu()方法中操作。
- 当菜单被选择的时候，在 OnOptionsItemSelected()方法中实现方法响应事件。

- 还可以调用 Menu 的 add()方法添加菜单项（MenuItem），可以调用 MenuItem 的 setIcon()方法为菜单项设置图标。

图 3-15 对应的代码实现如下：

```
1.  public class MenusActivity extends Activity {
2.      private static final int item1 = Menu.FIRST;
3.      private static final int item2 = Menu.FIRST + 1;
4.      public void onCreate(Bundle savedInstanceState) {
5.          super.onCreate(savedInstanceState);
6.          setContentView(R.layout.main);
7.      }
8.       public boolean onCreateOptionsMenu(Menu menu) {
9.          menu.add(0, item1, 0, "").setIcon(R.drawable.pic);
10.         menu.add(0, item2, 0, "上传");
11.         return true;
12.     }
13.     public boolean onOptionsItemSelected(MenuItem item) {
14.         switch (item.getItemId()) {
15.         case item1:
16.           setTitle("单击了菜单子项1");
17.           break;
18.         case item2:
19.           setTitle("单击了菜单子项2");
20.           break;
21.         }
22.      return true;
23.     }
24. }
```

详细代码见本书配套资源中的工程 Chapter3.4.1。

3.4.2 上下文菜单

上下文菜单（Context Menu）和 Windows 中的右键快捷菜单差不多。它一般是通过长按屏幕调用注册的上下文菜单。

如图 3-16 所示，程序运行后长按屏幕就会出现图中菜单。

实现上下文菜单要做的有以下几点：

- 覆盖 Activity 的 onCreateContextMenu()方法，调用 Menu 的 add()方法可以添加菜单项 MenuItem。
- 覆盖 onContextItemSelected()方法，响应菜单单击事件。
- 调用 registerForContextMenu()方法，为视图注册上下文菜单。

图 3-16 上下文菜单

图 3-16 对应的代码实现如下：

```
1.  public class MenusActivity extends Activity {
2.     private LinearLayout bc;
3.     public void onCreate(Bundle savedInstanceState) {
4.        super.onCreate(savedInstanceState);
5.        setContentView(R.layout.main);
6.        bc=(LinearLayout)findViewById(R.id.lay);
7.        registerForContextMenu(bc);
8.     }
9.     public void onCreateContextMenu(ContextMenu menu,View v,ContextMenuInfo
                                      menuInfo)
10.    {
11.       setTitle("菜单");
12.       menu.add(0,2,0,"菜单1");
13.       menu.add(0,3,0,"菜单2");
14.       super.onCreateContextMenu(menu, v, menuInfo);
15.    }
16.    public boolean onContextItemSelected(MenuItem item)
17.    {
18.       switch(item.getItemId()){
```

```
19.        case 2:
20.            Toast.makeText(this, "1", Toast.LENGTH_LONG).show();
21.            break;
22.        case 3:
23.            Toast.makeText(this, "2", Toast.LENGTH_LONG).show();
24.            break;
25.        }
26.        return true;
27.    }
28. }
```

详细代码见本书配套资源中的工程 Chapter3.4.2。

3.4.3 子菜单

子菜单（Submenu）就是将相同功能的分组进行多级显示的一种菜单，比如 Windows 的"文件"菜单中就有"新建""打开""关闭"等子菜单。

如图 3-17 所示，单击图中的任意一个菜单按钮就会出现图 3-18 中的子菜单。

图 3-17 主菜单

图 3-18 子菜单

有关子菜单需要注意以下几个方面：
- 通过触摸 Menu Item，调用子菜单选项。子菜单不支持嵌套，即子菜单中不能再包括其他子菜单。
- 覆盖 Activity 的 onCreateOptionsMenu()方法，调用 Menu 的 addSubMenu()方法添加子菜单项。
- 调用 SubMenu 的 add()方法，添加子菜单项。
- 覆盖 onCreateItemSelected()方法，响应菜单单击事件。

代码如下：

```
1.  public class MenusActivity extends Activity {
2.      public void onCreate(Bundle savedInstanceState) {
3.          super.onCreate(savedInstanceState);
4.          setContentView(R.layout.main);
5.      }
6.      public boolean onCreateOptionsMenu(Menu menu) {
7.          super.onCreateOptionsMenu(menu);
8.          SubMenu fileMenu=menu.addSubMenu(1, 1, 1, "File");
9.          SubMenu editMenu=menu.addSubMenu(1, 2, 2, "edit");
10.         fileMenu.add(2, 11, 11, "New");
11.         fileMenu.add(2, 12, 12, "Save");
12.         fileMenu.add(2, 13, 13, "Close");
13.         editMenu.add(2, 21, 21, "first");
14.         editMenu.add(2, 22, 22, "second");
15.         return true;
16.     }
17.     public boolean onOptionsItemSelected(MenuItem item) {
18.         super.onOptionsItemSelected(item);
19.         switch(item.getItemId()){
20.             case 1:{
21.                 Toast.makeText(MenusActivity.this,"单击了"+item.getTitle(),
                            Toast.LENGTH_SHORT).show();
22.                 break;
23.             }
24.             case 2:{
25.                 Toast.makeText(MenusActivity.this, "单击了"+item.getTitle(),
                            Toast.LENGTH_SHORT).show();
26.                 break;
27.             }
28.             case 11:{
29.                 Toast.makeText(MenusActivity.this, "单击了"+item.getTitle(),
                            Toast.LENGTH_SHORT).show();
30.                 break;
```

```
31.          }
32.          case 12:{
33.             Toast.makeText(MenusActivity.this, "单击了"+item.getTitle(),
                                Toast.LENGTH_SHORT).show();
34.              break;
35.          }
36.          case 13:{
37.             Toast.makeText(MenusActivity.this, "单击了"+item.getTitle(),
                                Toast.LENGTH_SHORT).show();
38.              break;
39.          }
40.          case 21:{
41.             Toast.makeText(MenusActivity.this, "单击了"+item.getTitle(),
                                Toast.LENGTH_SHORT).show();
42.              break;
43.          }
44.          case 22:{
45.             Toast.makeText(MenusActivity.this, "单击了"+item.getTitle(),
                                Toast.LENGTH_SHORT).show();
46.              break;
47.          }
48.      }
49.      return true;
50.  }
51. }
```

详细代码见本书配套资源中的工程Chapter3.4.3。

3.4.4 定义XML菜单文件

在此之前我们都是直接在代码中添加菜单项、给菜单分组等。在代码中添加菜单项是存在很多不足的，比如为了响应每个菜单项，就要用常量来保存每个菜单的 ID 等。那么有什么更好的方法来添加菜单呢？其实在 Android 中，可以把 menu 也定义为应用程序资源（也就是定义XML菜单文件），通过Android对资源的本地支持，可以更方便地实现菜单的创建与响应。

如图3-19所示，单击first1菜单按钮，出现图3-20中的子菜单。

创建步骤如下：

（1）在/res文件夹下创建menu文件夹。

（2）在menu文件夹下使用与menu相关的元素定义XML文件。

（3）使用XML文件的资源ID，将XML文件中定义的菜单项添加到menu对象中。

（4）响应菜单，使用每个菜单项所对应的资源ID。

图 3-19 主菜单　　　　　　　　　图 3-20 子菜单

在菜单的 XML 文件中，如果要创建子菜单，在 item 元素下嵌套一个 menu 就可以实现。

代码如下：

（1）XML 代码。

```
1.   <?xml version="1.0" encoding="utf-8"?>
2.   <menu xmlns:android="http://schemas.android.com/apk/res/android" >
3.     <item android:id="@+id/item1"
4.         android:title="first1">
5.       <menu >
6.         <item android:id="@+id/item5"
7.             android:title="one"/>
8.         <item android:id="@+id/item6"
9.             android:title="two"/>
10.      </menu>
11.    </item>
12.    <item android:id="@+id/item2"
```

```
13.            android:title="second2"/>>
14.        <item android:id="@+id/item3"
15.            android:title="first3"/>
16.        <item android:id="@+id/item4"
17.            android:title="second4"/>
18.    </menu>
```

（2）Java 代码。

```
1.  public class MenusActivity extends Activity {
2.          public void onCreate(Bundle savedInstanceState){
3.              super.onCreate(savedInstanceState);
4.              setContentView(R.layout.main);
5.          }
6.  public boolean onCreateOptionsMenu(Menu menu){
7.          MenuInflater inflater=getMenuInflater();
8.          inflater.inflate(R.menu.menu1, menu);
9.          return true;
10.         }
11. }
```

详细代码见本书配套资源中的工程 Chapter3.4.4。

3.5 事件响应

事件处理在 Android 开发中是一个非常重要的课题。事件是用户与界面交互时所触发的操作，比如单击一个按钮就会触发一个按钮的单击事件。事件处理是应用程序与用户交互的必要环节。在 Android 框架的设计中，以事件监听器（event listener）的方式来处理 UI 的使用者事件。

3.5.1 基本事件

在之前 Button 的介绍中已经讲到了如何为 Button 设置监听事件。我们已经知道，View 是绘制 UI 的基础类别，每个 View 组件都可以向 Android 框架注册一个事件监听器。每个事件监听器都包含一个回调函数（callback method），这个回调函数的作用就是处理用户的操作。Android 中常见的事件有以下几种：

- onClick(View v) 一个普通的单击按钮事件。
- boolean onKeyMultiple(int keyCode,int repeatCount,KeyEvent event)用于在多个事件连续出现时发生，重复按键，必须通过重载实现。
- boolean onKeyDown（int keyCode,KeyEvent event）在按键按下时发生。
- boolean onKeyUp（int keyCode,KeyEvent event）在按键释放时发生。
- onTouchEvent（MotionEvent event）触摸屏事件，当在触摸屏上有动作时发生。

- boolean onKeyLongPress（int keyCode, KeyEvent event）当长时间按下时发生。

上面这些事件都是在事件监听器中进行的，当需要响应某个组件的某个事件动作时，先要为该组件注册一个该事件的监听器。

3.5.2 事件的响应

下面就来完成图 3-21 的单击效果。

图 3-21 按钮单击

一个事件响应可通过 3 种方式实现。

首先是 XML 属性的方式，XML 布局代码如下：

```
1.  <LinearLayout xmlns:android="http://schemas.android.com/apk/res/android"
2.      android:layout_width="fill_parent"
3.      android:layout_height="fill_parent"
4.      android:orientation="vertical" >
5.      <Button
6.          android:id="@+id/button1"
7.          android:layout_width="wrap_content"
8.          android:layout_height="wrap_content"
9.          android:onClick="click"
10.         android:text="Button"/>
11. </LinearLayout>
```

然后需要在 Activity 中写一个 click 函数作为 Button 的响应函数。代码如下：

```
1.  public class MainActivity extends Activity {
2.      public void onCreate(Bundle savedInstanceState) {
3.          super.onCreate(savedInstanceState);
4.          setContentView(R.layout.main);
5.
6.      }
7.
8.      public void click(View v)
9.      {
10. Toast.makeText(MainActivity.this, "单击了按钮" + v.getId()
    Toast.LENGTH_SHORT).show();
11.     }
12. }
```

代码第 8、9 行就是按钮的响应函数具体实现，需要注意的是，这个响应函数应写在调用这个布局文件的 Activity 类中，而且函数的参数必须按照标准的响应函数的参数格式。运行后的效果如图 3-14 所示。我们刚才实现的代码见本书配套资源中的工程 Chapter3.5.1。

再来看一下让 Activity 实现监听器完成事件响应，代码如下：

```
1.  public class MainActivity extends Activity implements OnClickListener {
2.      public void onCreate(Bundle savedInstanceState) {
3.          super.onCreate(savedInstanceState);
4.          setContentView(R.layout.main);
5.      }
6.      public void onClick(View v) {
7.
8.          if (v.getId() == R.id.button1)
9.              Toast.makeText(MainActivity.this, "单击了按钮"+v.getId(),
10.                 Toast.LENGTH_SHORT).show();
11.     }
12. }
```

注意看第 1 行，MainActivity 除了继承了 Activity 还继承了 OnClickListener 接口并重载了 onClick()方法，在 onClick()方法中可以根据 View 的 Id 做出不同的响应，其运行效果也如图 3-14 所示。刚才实现的代码见本书配套资源中的工程 Chapter3.5.2。

最后以内部类的方式为按钮添加一个监听器完成事件响应。代码如下：

```
1.  public class MainActivity extends Activity {
2.      public void onCreate(Bundle savedInstanceState) {
3.          super.onCreate(savedInstanceState);
4.          setContentView(R.layout.main);
```

```
5.         Button btn = (Button)findViewById(R.id.button1);
6.         btn.setOnClickListener(new OnBtnClick());
7.     }
8.     private class OnBtnClick implements OnClickListener {
9.         public void onClick(View v) {
10.            if (v.getId() == R.id.button1)
11.                Toast.makeText(MainActivity.this, "单击了按钮"+v.getId(),
12.                    Toast.LENGTH_SHORT).show();
13.        }
14.    }
15. }
```

刚才实现的代码见本书配套资源中的工程 Chapter3.5.3。

至此，Android 的 3 种事件响应方式就都介绍完了，虽然只以 Button 的 OnClick()事件为例，但无论什么事件其原理都是一样的。读者可以自己研究一下其他事件的处理。

3.6 界面切换与数据传递

3.6.1 Intent 与 Bundle

Intent 与 Bundle 在界面切换中用得比较多。因为两个界面的切换肯定需要交代一些信息，这时就有 Intent 和 Bundle 的用武之地了，当然 Intent 的作用在于实现界面的切换。下面先来初步认识一下 Intent 和 Bundle，它们在数据传递中的用法会在后面中讲到。

1．Intent

我们已经知道 Android 有四大组件，这四大组件是独立的，它们之间相互协调，最终构成一个完整的 Android 应用。这些组件间的通信都是通过 Intent 的协助来完成的，Intent 负责对一次动作、动作涉及的数据进行描述，Android 根据这个 Intent 的描述找到相应组件，将 Intent 传给该组件并完成组件的调用。作为初学者需要理解的是 Intent 的两种启动模式。

- 显式的 Intent：即在构造 Intent 对象时就指定接收者，这种方式与普通的函数调用类似。
- 隐式的 Intent：即 Intent 的发送者在构造 Intent 对象时，并不知道也不关心接收者是谁。

1）显式的 Intent

在同一个程序中需要从当前 Activity 跳转到另一个指定的 Activity 时，就常用到显式的 Intent。

要创建一个显式的 Intent 可以使用构造函数 Intent(Context packageContext, Class<?> cls)。两个参数分别指定 Context 和 Class，Context 设置为当前的 Activity 对象，Class 设置为需要跳转到的 Activity 的类对象，比如要从当前界面跳到 TestActivity，就可以这样构造一个 Intent 对象：

```
1. Intent intent = new Intent(this, TestActivity.class);
2. startActivity(intent);
```

最后调用当前 Activity 的 startActivity()方法启动这个 Intent。除此之外，还可以采用下面的方法：

```
1. Intent intent=new Intent();
2. intent.setClass(this, TestActivity.class);
3. startActivity(intent);
```

这段代码采用了 setClass()方法将第一段代码中的第 1 行写成了两行，但目的是一样的。但这里使用的 Activity 都必须在 AndroidManifest.xml 文件中配置。

2）隐式的 Intent

其实，Intent 机制更重要的作用在于隐式的 Intent，即 Intent 的发送者不指定接收者，很可能不知道也不关心接收者是谁，而由 Android 框架去寻找最匹配的接收者。比如，程序要调用 Android 的电话功能，就要采用下面这种方式创建一个 Intent。代码如下：

```
1. Intent intent = new Intent(Intent.ACTION_DIAL);
2. startActivity(intent);
```

代码第 1 行创建 Intent 采用 Intent(String action)的构造函数，Action 可以理解为描述这个 Intent 的一种方式，Intent 的发送者只要指定了 Action 为 Intent.ACTION_DIAL，系统就能找到对应的 Activity 作为接收者。Android 系统提供了很多的 Action，读者可以查看官方的 API 文档（reference\android\content\Intent）。同样也可以使用 setAction(String action)的方法来达到同样的效果。代码如下：

```
1. Intent intent = new Intent();
2. intent. setAction (Intent.ACTION_DIAL);
3. startActivity(intent);
```

不管是显式还是隐式的 Intent，在完成 Activity 的切换时都可能涉及数据的传递，Intent 提供了一系列的方法，如 putExtra(String name, String value)：采用 Key-Value 的形式，Key 是数据的键，Value 是数据的值。该方法在 Intent 中有多种重载形式，读者可以查看 Intent 的 API。有关数据传递部分将在后面详细讲解。

2. Bundle

Bundle 其实就是一个 Key-Value 的映射，前面介绍 Intent 时说到，Intent 描述数据的 putExtra 方法采用的是 Ket-Value 形式，其实 Intent 在内部定义时就有个 Bundle 类型的成员变量，putExtra 方法就是将传进来的参数放到这个 Bundle 变量中。所以 Bundle 有许多类似于 putExtra 的方法来存放数据，这里就不多讲解了。这里介绍 Bundle 用于传递对象时的几个方法。

首先要传递对象不能是一般的对象，这个对象必须是可序列化的，这就要求它们继承 Parcelable 或 Serializable 接口。

- putParcelable (String key, Parcelable value)：顾名思义，该方法用于将 Parcelable 接口的对象存入 Bundle 中，对应的 getParcelable (String key)用于接收键值为 key 的 Parcelable 对象。
- putSerializable(String key, Serializablevalue)：与 putParcelable 的用法一样，只不过该方法用于存放实现了 Serializable 接口的对象。

Intent 和 Bundle 就先了解到这里，有关它们的其他知识会在后面涉及，也可查看官方 API 进行更深入的了解。

3.6.2 界面切换

在一个程序中，经常需要将当前的 Activity 转跳到另一个 Activity 进行操作，这时就需要运用到 Activity 切换。Activity 切换有两种方式，下面就用这两种方式来实现从图 3-22 的界面切换到图 3-23 的界面。

图 3-22　MainActivity 界面

图 3-23　LoginActivity 界面

方法一，可以通过 setContentView 切换布局来实现界面的切换。步骤如下：

（1）新建一个想要切换的界面的 XML 文件。

（2）通过触发一个加载了监听器的控件，在监听器中使用 setContentView 函数切换界面。这样的实现过程都是在一个 Activity 上面实现的，所有变量都在同一状态，因此

所有变量都可以在这个 Activity 状态中获得。这里可以设置一个加载了监听器的 Button 按钮,实现代码如下:

```
1.  public void onCreate(Bundle savedInstanceState) {
2.      super.onCreate(savedInstanceState);
3.      setContentView(R.layout.main);
4.      Button button = (Button) this.findViewById(R.id.button1);
5.        button.setOnClickListener(new OnClickListener() {
6.      public void onClick(View v) {
7.        setContentView(R.layout.login);
8.            }
9.          });
10.  }
```

这样就从 main 这个 XML 布局切换到了 login 这个 XML 布局。但严格意义上讲,这并不是一种界面的切换而是对界面的重画,就像是老师上课的时候将黑板上以前的板书擦掉重新写一块板书。

☞**注意**:在有些教程中会看到方法一,但我们不提倡使用这种方法,应使用方法二。

刚才实现的代码见本书配套资源中的工程 Chapter3.6.1。

方法二,在一个程序中往往会使用 Intent 对象来指定一个 Activity,并通过 startActivity 方法启动这个 Activity。当需要在不同的 Activity 之间进行切换的时候,可以在响应事件中实例化一个 Intent 对象作为 startActivity 方法的参数,从而实现不同 Activity 的切换。

下面完成相同的界面切换,我们需要做的是将方法一代码中的第 9 行改成下面这段代码:

```
1.  Intent intent=new Intent();
2.      intent.setClass(MainActivity.this, LoginActivity.class);
3.      MainActivity.this.startActivity(intent);
```

上述代码实现了从当前 MainActivity 切换到 LoginActivity。

刚才实现的代码见本书配套资源中的工程 Chapter3.6.2。

3.6.3 传递数据

1. Intent 数据传递

之前已经介绍了 Android 利用 Intent 完成 Activity 间的切换。在界面切换的时候经常需要数据的传递,同样需要用到 Intent。我们先来完成两个简单的 Activity 间的数据传递。首先需要两个 Activity:DataTransferActivity 和 ShowActivity,界面效果如图 3-24 所示。

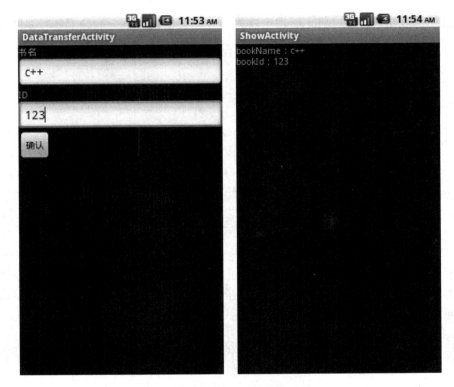

图 3-24　Activity 间的数据传递

当单击了 DataTransferActivity 中的"确认"按钮时就会切换到 ShowActivity 界面，同时将编辑框中的书名和 ID 传递到 ShowActivity 中显示出来。下面主要研究数据传递的代码。下面是事件响应代码。

```
1.    Button btn = (Button) findViewById(R.id.button1);
2.          btn.setOnClickListener(new OnClickListener() {
3.             public void onClick(View v) {
4.                EditText bookNameTV = (EditText) findViewById(R.id.
                         editText1);
5.                EditText bookIdTV = (EditText) findViewById(R.id.
                         editText2);
6.                Intent intent = new Intent(DataTransferActivity.this,
                         ShowActivity.class);
7.                intent.putExtra("bookName",bookNameTV.getText().
                         toString());
8.                intent.putExtra("bookId", bookIdTV.getText().toString());
9.                DataTransferActivity.this.startActivity(intent);
10.            }
11.        });
```

注意第 7、8 行代码，这里调用了 Intent 的 putExtra 方法传递 bookName 和 bookId，

而且它们每一个都有一个键值，这是为了方便获取时使用。putExtra 方法在 Intent 中有多种重载形式，可以存放多种类型的数据。那么接下来再看一下 ShowActivity 是怎样接收数据的。代码如下：

```
1.   public void onCreate(Bundle savedInstanceState) {
2.       super.onCreate(savedInstanceState);
3.       Intent intent = getIntent();
4.       String bookName = intent.getStringExtra("bookName");
5.       String bookId =  intent.getStringExtra("bookId");
6.       TextView textview = new TextView(this);
7.       textview.setText("bookName: "+bookName+"\nbookId: "+bookId);
8.       setContentView(textview);}
```

上面的代码中第 4、5 行的作用就是接收数据。它调用了 Intent 的 getStringExtra 方法，用于获取 String 类型的数据，其他数据类型获取方法也与此类似，就是通过键值来获取对应的数据。但是有时候数据太多，类型也不一样，使用这样的方法就有些复杂了，下面介绍另一种数据传递的方式——利用 Bundle 传递数据。

2．Bundle 传递对象

查看 Intent 的源码可以发现，其实 Intent 的数据传递同样是将数据绑定在 Bundle 中传递的。

Bundle 还有一个功能就是传递对象，但前提是这个对象需要序列化。还是如图 2-9 所示的效果，但这次将 bookName 和 bookId 封装在一个 Book 对象中，然后利用 Bundle 和 IntentBook 对象传递到 ShowActivity 中。首先看一下 Book 类的代码。代码如下：

```
1.   public class Book implements Serializable{
2.       public String getBookName() {
3.           return bookName;
4.       }
5.       public String getBookId() {
6.           return bookId;
7.       }
8.       private String bookName = null;
9.       private String bookId = null;
10.
11.      public Book(String bookName,String bookId)
12.      {
13.          this.bookId = bookId;
14.          this.bookName = bookName;
15.      }
16.  }
```

因为需要 Book 的对象是可序列化的，所以第 1 行代码让 Book 继承了 Serializable 接口。再看一下 DataTransferActivity 中事件监听改变的代码：

```java
1.  btn.setOnClickListener(new OnClickListener() {
2.
3.          public void onClick(View v) {
4.              EditText bookNameTV = (EditText) findViewById(R.id.
                                  editText1);
5.              EditText bookIdTV = (EditText) findViewById(R.id.
                                  editText2);
6.              Intent intent = new Intent(DataTransferActivity.this,
                                  ShowActivity.class);
7.              Bundle mExtra = new Bundle();
8.              String bookName = bookNameTV.getText().toString();
9.              String bookId = bookIdTV.getText().toString();
10.             Book book= new Book(bookName,bookId);
11.             mExtra.putSerializable("book", book);
12.             intent.putExtras(mExtra);
13.             DataTransferActivity.this.startActivity(intent);
14.
15.         }
16.     });
```

上面代码的第 7 行定义了一个 Bundle 对象 mExtra；第 10 行代码将 bookName 和 bookId 封装到旧 book 对象中，因为 Book 实现的是 Serializable 接口所以调用 Bundle 中对应的 putSerializable 方法，参数同样采用了 Key-Value 的形式；最后调用 intent 的 putExtras 方法将 mExtra 存放到 intent 中。这样剩下的就是如何接收了。下面就来看下 ShowActivity 是怎样接收的。代码如下：

```java
1.  public void onCreate(Bundle savedInstanceState) {
2.      super.onCreate(savedInstanceState);
3.      Intent intent = getIntent();
4.      Book book = (Book) intent.getSerializableExtra("book");
5.      String bookName = book.getBookName();
6.      String bookId = book.getBookId();
7.      TextView textview = new TextView(this);
8.      textview.setText("bookName: "+bookName+"\nbookId: "+bookId);
9.      setContentView(textview);
10. }
```

既然传递时存放对象采用的是 putSerializable 方法，接收时肯定采用对应的 getSerializableExtra 方法，参数就是这个对象的键值，然后再将之强制转换为 Book 对象，并调用 Book 的相应方法获得想要的数据。

这里使用的继承 Serializable 接口序列化对象，还可以实现 Parcelable 接口，但对应的存放和读取对象方式就改为 putParcelable 和 getParcelable 方法。

本部分代码见本书配套资源中的工程 Chapter3.6.3。

3.7 Activity 界面刷新

在进行 Activity 刷新时不能在子线程中进行,只能在该程序的主线程中刷新 Activity。当需要在子线程中进行刷新时,主线程运行子线程,然后从子线程中返回一个刷新的操作对主线程进行刷新,如图 3-25 所示。

图 3-25 界面刷新

3.8 Activity 栈及 4 种启动模式

3.8.1 Activity 栈概述

1. Activity 栈

开发者是无法控制 Activity 的状态的,那么 Activity 的状态又是按照何种逻辑来运作的呢?这就要知道 Activity 栈。

每个 Activity 的状态是由它在 Activity 栈(包含所有正在运行 Activity 的队列)中的位置决定的。

当一个新的 Activity 启动时,当前活动的 Activity 将会移到 Activity 栈的顶部。

如果用户使用后退按钮返回,或者前台的 Activity 结束,在栈上的 Activity 将会移上来并变为活动状态。

一个应用程序的优先级是受最高优先级的 Activity 影响的。当决定某个应用程序是否要终结去释放资源,Android 内存管理使用栈来决定基于 Activity 的应用程序的优先级。

2. Activity 的 4 种状态

1)活动的

当一个 Activity 在栈顶时,它是可视、有焦点、可接收用户输入的。Android 试图尽最大可能保持其活动状态,杀死其他 Activity 来确保当前活动 Activity 有足够的资源可以使用。当另一个 Activity 被激活时,这个将会被暂停。

2)暂停

在很多情况下,某个 Activity 可视但是它没有焦点,换句话说它被暂停了。可能的原因是一个透明或者非全屏的 Activity 被激活。

当被暂停时,一个 Activity 仍会处于活动状态,只不过是不可以接收用户输入。在极特殊的情况下,Android 将会杀死一个暂停的 Activity 来为活动的 Activity 提供充足的资源。当一个 Activity 变为完全隐藏的时候,它将会变成停止。

3）停止

当一个 Activity 不是可视的，它就是"停止"的。这个 Activity 将仍然在内存中保存其所有的状态和会员信息。尽管如此，当其他地方需要内存时，它将是最有可能被释放资源的。当一个 Activity 停止后，一个很重要的步骤是要保存数据和当前 UI 状态。一旦一个 Activity 退出或关闭了，它将变为待用状态。

4）待用

在一个 Activity 被杀死后和被装载前，它是待用状态的。待用 Acitivity 被移出 Activity 栈，并且需要在显示和可用之前重新启动它。

3.8.2 Activity 启动模式定义方法

在 Android 的多 Activity 开发中，Activity 之间的跳转可能需要有多种方式，有时是普通地生成一个新实例，有时希望跳转到原来某个 Activity 实例，而不是生成大量重复的 Activity。加载模式便是决定以哪种方式启动一个或跳转到原来某个 Activity 实例。

启动模式有两种不同的定义方法。

1）使用清单文件

当在清单文件（AndroidManifest.xml）中声明一个 Activity 时，你能够指定这个 Activity 在启动时应该如何与任务进行关联。这些启动模式可以在功能清单文件 AndroidManifest.xml 中设置 launchMode 属性。

2）使用 Intent 标识

在调用 startActivity()方法时，你能够在 Intent 中包含一个标识，用来声明这个新的 Activity 应该如何与当前的任务进行关联。

Intent 标识设置实例如下：

```
1.   Intent intent = new Intent(MainActivity.this, SecActivity.class);
2.   intent.addFlags(Intent.FLAG_ACTIVITY_REORDER_TO_FRONT);
3.   startActivity(intent);
```

下面列举一些常用的 Intent 标识：

- FLAG_ACTIVITY_BROUGHT_TO_FRONT——这个标识一般不是由程序代码设置的，如在 launchMode 中设置 singleTask 模式时系统帮你设定。
- FLAG_ACTIVITY_CLEAR_TOP——当这个 Activity 已经在当前的 Task 中运行时，Android 不再重新启动一个这个 Activity 的实例，而是在这个 Activity 上方的所有 Activity 都将关闭，然后这个 Intent 会作为一个新的 Intent 投递到旧的 Activity（现在位于顶端）中。
- FLAG_ACTIVITY_CLEAR_WHEN_TASK_RESET——若设置了这个标识，则将在 Task 的 Activity 栈中设置一个还原点，当 Task 恢复时，需要清理 Activity。也就是说，下一次 Task 带着 FLAG_ACTIVITY_RESET_TASK_IF_NEEDE 标识进入前台时（典型的操作是用户在主画面重启它），这个 Activity 和它之上的 Activity 都将关闭，以至于用户不能再返回到它们，但是可以回到之前的 Activity。

- FLAG_ACTIVITY_EXCLUDE_FROM_RECENTS——若设置了这个标识,则新的 Activity 不会在最近启动的 Activity 的列表中保存。
- FLAG_ACTIVITY_FORWARD_RESULT——如果这个 Intent 用于从一个存在的 Activity 启动一个新的 Activity,那么作为答复目标的 Activity 将会传到新的 Activity 中。在这种方式下,新的 Activity 可以调用 setResult(int),并且这个结果值将发送给那个作为答复目标的 Activity。

除此以外,Intent 还有很多的标识,无法全部列举,读者可以查看 Intent 的 API 了解更多的标识。

3.8.3 standard 启动模式

这是默认模式,每次激活 Activity 时都会调用 startActivity()方法创建一个新的 Activity 实例,并放入任务栈中。

假如此时有 A、B 两个 Activity,如果 Activity A 采用 standard 模式启动 Activity B,则不管 Activity B 是否已经有一个实例位于 Activity 栈中,都会产生一个 Activity B 的新实例并放入任务栈中。

清单文件配置方式:

```
1.   <activity
2.       android:name=".StandardActivity"
3.       android:label="@string/standard"
4.       android:launchMode="standard">
5.   </activity>
```

standard 的运行机制,部分代码如下:

```
1.   private TextView text;
2.       private Button button;
3.           public void onCreate(Bundle savedInstanceState) {
4.               super.onCreate(savedInstanceState);
5.               setContentView(R.layout. mainactivity);
6.               text = (TextView) this.findViewById(R.id.text);
7.               text.setText(this.toString());
8.               button = (Button) this.findViewById(R.id.button_stand);
9.           }
10.      //按钮单击事件
11.          public void LaunchStandard(){
12.              startActivity(new Intent(this,StandardActivity.class));
13.              text.setText(this.toString());
14.          }
```

初始化界面如图 3-26 所示。

当单击按钮时,会创建新的 Activity,通过 TextView@后的十六进制数的显示可看出界面如图 3-27 所示。

图 3-26 standard 模式的初始界面

图 3-27 单击按钮后的界面

再次单击按钮时，还会创建新的 Activity，通过 TextView@后的十六进制数的显示可看出界面如图 3-28 所示。

分析其运行机制可知，当程序运行到此时，栈中的数据形式如图 3-29 所示。

图 3-28 再次单击按钮后的界面

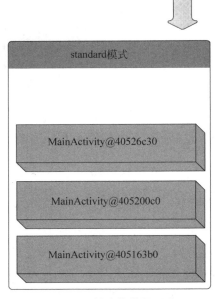
图 3-29 栈中的数据形式

因此，这种 standard 模式是每次都会创建新的 Activity 对象，当单击返回按钮时，它会将栈顶（当前 Activity）消灭，然后跳到下一层，例如，如果现在 Activity 是 405263c0，那么当单击返回时 Activity 会变为 405200c0，不过此时在这个 Activity 中再次单击按钮创建对象时，它会另外创建新的 Activity 对象，这种模式可能在大多数情况下不是我们需要的，因为对系统性能的消耗过大。

完整代码见工程 Chapter3.8.3。

3.8.4　singleTop 启动模式

如果已经有一个实例位于 Activity 栈的顶部，就不产生新的实例，而只是调用 Activity 中的 newInstance() 方法。如果不位于栈顶，则会产生一个新的实例。

清单文件配置方式：

```
1.   <activity
2.        android:name=".SingleTopActivity"
3.        android:label="@string/singleTop"
4.        android:launchMode="singleTop" >
5.   </activity>
```

界面初始化如图 3-30 所示。

单击"启动 singleTop 模式"按钮，结果如图 3-31 所示。

图 3-30　singleTop 模式的初始界面　　图 3-31　单击"启动 singleTop 模式"按钮后的界面

分析其运行机制可知,当程序运行到此时,栈中的数据形式如图 3-32 所示。

再次单击"启动 singleTop 模式"按钮,结果如图 3-33 所示,由于此 Activity 设置的启动模式为 singleTop,因此它首先会检测当前栈顶是否为我们要请求的 Activity 对象,经验证成立,因此它不会创建新的 Activity,而是引用当前栈顶的 Activity。

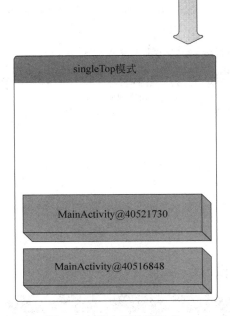

图 3-32　栈中的数据形式

图 3-33　再次单击"启动 singleTop 模式"按钮

此时栈中的数据形式依然为图 3-32。

完整代码见工程 Chapter3.8.4。

3.8.5　singleTask 启动模式

如果在栈中已经有该 Activity 的实例,则重用该实例(会调用实例的 onNewIntent())。重用时,会让该实例回到栈顶,因此在它上面的实例将会被移出栈。如果栈中不存在该实例,将会创建新的实例放入栈中。

清单文件配置方式:

```
1.  <activity
2.      android:name=".SingleTaskActivity"
3.      android:label="@string/singleTask"
4.      android:launchMode="singleTask" >
5.  </activity>
6.
```

界面初始化为如图 3-34 所示的形式。

单击"启动 singleTask 模式"按钮,界面如图 3-35 所示。

图 3-34　singleTask 模式初始化界面

图 3-35　单击"启动 singleTask 模式"按钮

在此界面中单击"启动 singleTask 模式"按钮，根据定义会检测当前栈中是否有此 Activity 对象，因此显示的还是当前的 Activity，不会重新创建，如图 3-36 所示。

再次单击"启动 standard 模式"按钮，由于 MainActivity 的启动模式为 standard，所以在此会重新创建一个 MainActivity 对象，如图 3-37 所示。

图 3-36　再次单击"启动 singleTask 模式"按钮

图 3-37　再次单击"启动 standard 模式"按钮

此时栈中数据形式如图 3-38 所示。

当在如图 3-39 所示界面中单击"启动 singleTask 模式"按钮时，由于检测到当前栈中第二个为我们要创建的 Activity，所以会将最上面的 MainActivity 消灭，然后将 SingleTaskActivity 设置为栈顶。

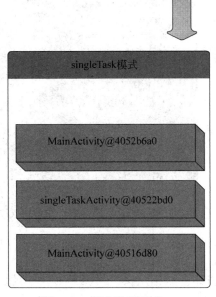
图 3-38　栈中数据形式

图 3-39　再次单击"启动 singleTask 模式"按钮

此时栈中数据形式变为如图 3-40 所示。

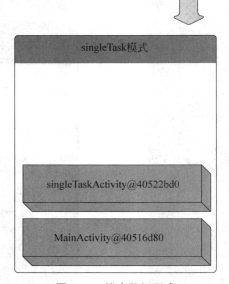
图 3-40　栈中数据形式

完整代码见工程 Chapter3.8.5。

3.9 有多个界面的单机版图书管理系统

本节将在 Chapter02 上进行改进,实现一个有多个界面的单机版图书管理系统。本节灵活运用了本章所学的 3 个知识点:界面组件、布局、事件监听和 Activity 之间的切换。首先在虚拟机上运行 Chapter03,效果如图 3-41 所示。

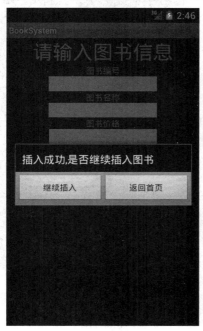

图 3-41　运行效果图

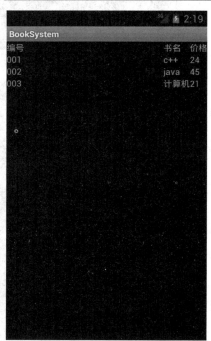

图 3-41（续）

在 Chapter03 中有 3 个包，分别是 ui.cqupt、control.cqupt 和 model.cqupt，如图 3-42 所示。

```
▲ 📁 Chapter03
    ▲ 🗂 src
        ▲ ⊞ control.cqupt
            ▷ 🗋 Controller.java
        ▲ ⊞ model.cqupt
            ▷ 🗋 Book.java
            ▷ 🗋 BookList.java
        ▲ ⊞ ui.cqupt
            ▷ 🗋 DeleteActivity.java
            ▷ 🗋 InsertActivity.java
            ▷ 🗋 MainActivity.java
            ▷ 🗋 SelectActivity.java
            ▷ 🗋 SetActivity.java
```

图 3-42　包结构

　　ui.cqupt 包的功能是对界面操作，包括了图 3-42 所示的所有界面的类。control.cqupt 包起到控制的作用，ui.cqupt 包通过 control.cqupt 包对 model.cqupt 包进行操控。model.cqupt 包用于存放模型，里面的 BookList 存储了图书的信息。这种界面、控制、模型的设计思路就是典型的 MVC 设计模式，MVC 设计模式的框架如图 3-43 所示。

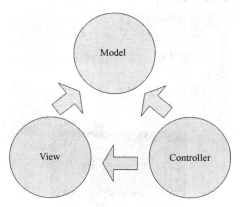

图 3-43　MVC 三层架构

1）什么是 MVC 设计模式

　　MVC 模式（三层架构模式）（Model-View-Controller）是软件工程中的一种软件架构模式，把软件系统分为 3 个基本部分：模型（Model）、视图（View）和控制器（Controller）。

　　控制器（Controller）——负责转发请求，对请求进行处理。

　　视图（View）——界面设计人员在此进行图形界面设计。

　　模型（Model）——程序员利用它编写程序应有的功能（实现算法等等）、数据库专家进行数据管理和数据库设计（可以实现具体的功能）。

2）为什么要使用 MVC 设计模式

MVC 实现了视图层和业务层分离，这样就允许更改视图层代码而不用重新编写模型和控制器代码；同样，一个应用的业务流程或者业务规则的改变只需要改动 MVC 的模型层即可，因为模型与控制器和视图相分离。这种分离有许多好处：

（1）清晰地将应用程序分隔为独立的部分；

（2）业务逻辑代码能够很方便地在多处重复使用；

（3）方便开发人员分工协作；

（4）如果需要，可以方便开发人员对应用程序各个部分的代码进行测试。

这里已经对 Chapter03 的整体结构和设计思路有了一定的了解。Chapter03 中各个类之间的关系如图 3-44 所示。

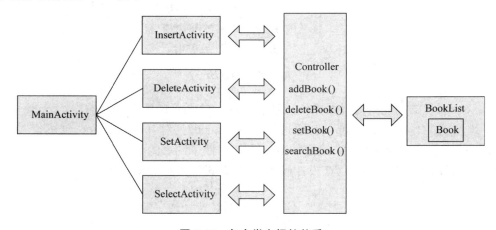

图 3-44　各个类之间的关系

下面主要对增加图书的流程做详细讲解，删除、修改和查询图书将在增加图书流程讲解之后介绍。

图 3-45　Chpater03 主界面

Chapter03 主界面如图 3-45 所示是由 ui.cqupt 包中的 MainActivity.java 实现的，运用

了图形界面和事件监听的知识点。下面将对主界面的实现进行详细的分析，MainActivity.java 代码如下所示：

```
1.  package ui.cqupt;
2.
3.  import ui.cqupt.R;
4.  import android.app.Activity;
5.  import android.content.Intent;
6.  import android.os.Bundle;
7.  import android.view.View;
8.  import android.view.View.OnClickListener;
9.  import android.widget.Button;
10.
11. public class MainActivity extends Activity {
12.
13.     public void onCreate(Bundle savedInstanceState) {
14.         super.onCreate(savedInstanceState);
15.         setContentView(R.layout.main);
16.         Button insert = (Button) findViewById(R.id.m_insert);
17.         Button delete = (Button) findViewById(R.id.m_delete);
18.         Button set = (Button) findViewById(R.id.m_set);
19.         Button select = (Button) findViewById(R.id.m_select);
20.         ButtonListener buttonListener = new ButtonListener();
21.         insert.setOnClickListener(buttonListener);
22.         delete.setOnClickListener(buttonListener);
23.         set.setOnClickListener(buttonListener);
24.         select.setOnClickListener(buttonListener);
25.     }
26.
27.     class ButtonListener implements OnClickListener {
28.
29.         public void onClick(View v) {
30.             int id = v.getId();
31.             Intent intent = new Intent();
32.             switch (id) {
33.             case R.id.m_insert:
34.                 intent.setClass(MainActivity.this, InsertActivity.class);
35.                 MainActivity.this.startActivity(intent);
36.                 break;
37.             case R.id.m_delete:
38.                 intent.setClass(MainActivity.this, DeleteActivity.class);
39.                 MainActivity.this.startActivity(intent);
40.                 break;
41.             case R.id.m_set:
42.                 intent.setClass(MainActivity.this, SetActivity.class);
43.                 MainActivity.this.startActivity(intent);
44.                 break;
45.             case R.id.m_select:
46.                 intent.setClass(MainActivity.this, SelectActivity.class);
```

```
47.                    MainActivity.this.startActivity(intent);
48.                    break;
49.             }
50.        }
51.
52.   }
53. }
```

MainActivity 首先覆写了 onCreate(Bundle savedInstanceState)方法，在方法体中第 15 行设置了 MainActivity 的界面布局——main.xml，布局文件 main.xml 代码如下所示：

```
1.  <?xml version="1.0" encoding="utf-8"?>
2.  <LinearLayout xmlns:android="http://schemas.android.com/apk/res/android"
3.      xmlns:tools="http://schemas.android.com/tools"
4.      android:layout_width="fill_parent"
5.      android:layout_height="fill_parent"
6.      android:orientation="vertical"
7.      tools:ignore="HardcodedText" >
8.      <TextView
9.          android:textSize="15pt"
10.         android:layout_width="fill_parent"
11.         android:layout_height="wrap_content"
12.         android:text="图书管理系统" />
13.     <Button
14.         android:id="@+id/m_insert"
15.         android:layout_width="wrap_content"
16.         android:layout_height="wrap_content"
17.         android:text="增" />
18.     <Button
19.         android:id="@+id/m_delete"
20.         android:layout_width="wrap_content"
21.         android:layout_height="wrap_content"
22.         android:text="删" />
23.     <Button
24.         android:id="@+id/m_set"
25.         android:layout_width="wrap_content"
26.         android:layout_height="wrap_content"
27.         android:text="改" />
28.     <Button
29.         android:id="@+id/m_select"
30.         android:layout_width="wrap_content"
31.         android:layout_height="wrap_content"
32.         android:text="查" />
33. </LinearLayout>
```

主界面的布局采用了线性布局，所有组件都放在<LinearLayout></LinearLayout>标签中，第 13～17 行的<Button/>标签为声明一个按钮组件，每个 Button 标签代表一个按钮组件，标签中的代码为定义这个组件的属性，第 14 行为此 Button 设定一个名为 m_insert 的 id，程序会在 R.java 中生成一个 id 名为 m_insert 的代码，此 id 唯一标识这个 Button 组件。第 17 行设定了 Button 组件上的显示，其余 3 个组件也以相同的方式设定。R.java 中的代码如下所示：

```
1.  public static final int m_delete=0x7f050007;
2.  public static final int m_insert=0x7f050006;
3.  public static final int m_select=0x7f050009;
4.  public static final int m_set=0x7f050008;
```

MainActivity.java 中第 16～19 行通过 findViewById(R.id.XX)方法查找 R.java 文件中生成名为 xx 的 id 号来实例化在布局文件中声明的按钮主件。第 20 行定义一个 Button 监听器对象；第 21～24 行，通过 Button 类的 setOnClickListener()方法设置处理单击事件的对象实例；第 27～52 行定义了一个 ButtonListener 的内部类来实现监听器，ButtonLiistener 实现了 android.view.View.OnClickListener 接口的 public void onClick(View v) 方法处理单击事件。

第 29～50 行实现了单击事件的处理，完成了 Activity 之间的切换功能。第 30 行得到按钮的唯一标识 id，并通过 switch 语句判断单击事件的执行代码。第 31 行定义了一个 intent 对象，Activity 通过 intend 对象完成界面的切换，第 34 行通过 Intend 的 setClass(MainActivity.this, InsertActivity.class)设置了从主界面跳转到 InsertActivity 插入界面，在第 35 行通过访问外部类的 startActivity(intent)方法启动 intent 实现界面切换，其余的 3 个组件也通过同样的方式实现了 Activity 的切换。4 个按钮组件执行后的结果如图 3-46～图 3-49 所示。

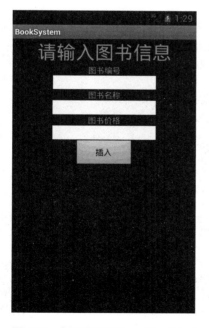

图 3-46　插入界面（InsertActivity）

图 3-47　删除界面（DeleteActivity）

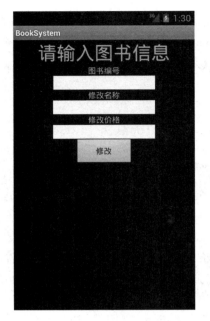 

图 3-48　修改界面（SetActivity）　　　图 3-49　查询界面（SelectActivity）

首先分析插入界面（insertActivity）。InsertActivity 位于 ui.cqupt 包中，代码如下所示：

```
1.  package ui.cqupt;
2.
3.  import control.cqupt.Controller;
4.  import ui.cqupt.R;
5.  import android.app.Activity;
6.  import android.app.AlertDialog.Builder;
7.  import android.content.DialogInterface;
8.  import android.os.Bundle;
9.  import android.view.View;
10. import android.view.View.OnClickListener;
11. import android.widget.Button;
12. import android.widget.EditText;
13.
14. public class InsertActivity extends Activity {
15.     private EditText name;
16.     private EditText id;
17.     private EditText price;
18.
19.     public void onCreate(Bundle savedInstanceState) {
20.         super.onCreate(savedInstanceState);
21.         setContentView(R.layout.insert);
22.         name = (EditText) findViewById(R.id.name);
```

```
23.         id = (EditText) findViewById(R.id.id);
24.         price = (EditText) findViewById(R.id.price);
25.         Button insert = (Button) findViewById(R.id.i_insert);
26.         insert.setOnClickListener(new ButtonListener());
27.     }
28.
29.     class ButtonListener implements OnClickListener {
30.
31.         public void onClick(View v) {
32.             String bookname = name.getText().toString();
33.             String bookid = id.getText().toString();
34.             String bookprice = price.getText().toString();
35.
36.             Controller control = new Controller();
37.             if (bookname.equals("") || bookid.equals("")
38.                     || bookprice.equals("")) {
39.                 new Builder(InsertActivity.this).setMessage("图书信息不能为空").show();
40.             } else {
41.                 if (control.addBook(bookid, bookname, bookprice)) {
42.                     id.setText("");
43.                     name.setText("");
44.                     price.setText("");
45.                     buildDialog();
46.                 } else {
47.                     new Builder(InsertActivity.this).setMessage("已有此图书").show();
48.                 }
49.             }
50.         }
51.
52.     private void buildDialog() {
53.         Builder builder = new Builder(InsertActivity.this);
54.         builder.setTitle("插入成功,是否继续插入图书");
55.         builder.setNegativeButton("返回首页",
56.                 new DialogInterface.OnClickListener() {
57.                     public void onClick(DialogInterface dialog,
58.                             int whichButton) {
59.                         finish();
60.                     }
61.
62.                 });
63.         builder.setPositiveButton("继续插入", null);
64.         builder.show();
```

65.   }
66.
67.   }
68. }

InsertActivity 继承了 Activity 并且覆写了 onCreate()方法初始化界面。本界面的布局采用线性布局，使用了 Button、TextView 和 EditText 组件，布局文件 insert.xml 代码如下所示：

```
1.  <?xml version="1.0" encoding="utf-8"?>
2.  <LinearLayout xmlns:android="http://schemas.android.com/apk/res/android"
3.      xmlns:tools="http://schemas.android.com/tools"
4.      android:layout_width="fill_parent"
5.      android:layout_height="fill_parent"
6.      android:gravity="center|top"
7.      android:orientation="vertical"
8.      tools:ignore="HardcodedText" >
9.
10.     <TextView
11.         android:layout_width="fill_parent"
12.         android:layout_height="wrap_content"
13.         android:gravity="center_horizontal"
14.         android:text="请输入图书信息"
15.         android:textSize="15pt" />
16.
17.     <TextView
18.         android:layout_width="wrap_content"
19.         android:layout_height="wrap_content"
20.         android:text="图书编号" />
21.
22.     <EditText
23.         android:id="@+id/id"
24.         android:layout_width="180dp"
25.         android:layout_height="wrap_content"
26.         android:background="#FFFFFF"
27.         android:ems="10"
28.         android:inputType="text"
29.         android:textColor="#000000" />
30.
31.     <TextView
32.         android:layout_width="wrap_content"
33.         android:layout_height="wrap_content"
34.         android:text="图书名称" />
35.
```

```
36.     <EditText
37.         android:id="@+id/name"
38.         android:layout_width="180dp"
39.         android:layout_height="wrap_content"
40.         android:background="#FFFFFF"
41.         android:ems="10"
42.         android:inputType="text"
43.         android:textColor="#000000" />
44.
45.     <TextView
46.         android:layout_width="wrap_content"
47.         android:layout_height="wrap_content"
48.         android:text="图书价格" />
49.
50.     <EditText
51.         android:id="@+id/price"
52.         android:layout_width="180dp"
53.         android:layout_height="wrap_content"
54.         android:background="#FFFFFF"
55.         android:ems="10"
56.         android:inputType="text"
57.         android:textColor="#000000" />
58.
59.     <Button
60.         android:id="@+id/i_insert"
61.         android:layout_width="102dp"
62.         android:layout_height="wrap_content"
63.         android:text="插入" />
64.
65. </LinearLayout>
```

在 insert.xml 文件中，第 10～15 行通过 TextView 组件在界面上显示 "请输入图书信息" 的信息，第 22～29 行通过<EditText/>标签声明一个文本框，在文本框中可输入图书的信息。Insert.xml 文件中所有<EditText/>标签都是声明的一个文本框，通过 android:id="@+id/ 名称 " 在 R.java 文件中生成唯一标识 id。第 26 行 android:background="#FFFFFF"设置文本框背景颜色，第 27 行设置字体大小，第 28 行设置输入类型，第 29 行设置输入的字体颜色。第 59～63 行声明了一个 Button 组件，命名为 "插入"。

现在回到 InsertActivity.java 的代码中，第 15～17 行声明了 EditText 组件，分别命名为 id、name、price。在第 22～24 行通过 findViewById(R.id.xx)实例化 EditText 组件，第 25 行实例化 Button 组件，并且在第 26 行通过 Button 类的 setOnClickListener()方法设置处理单击事件的对象实例。第 29～67 行定义了一个 ButtonListener 的内部类来实现监听

器,并且实现了 public void onClick(View v)方法进行事件处理。第 32～34 行通过 getText()方法得到 EditText 文本框中的内容,并且使用 toString()方法转换为字符串。第 36 行创建一个 control 对象,通过 control 对象对存储数据的 BookList 操作。第 37～50 行是对事件处理的核心,首先在第 37、38 行判断输入框是否为空,如果为空,就生成一个消息对话框,显示 "图书信息不能为空"。第 41～48 行通过调用 control 的 addBook 方法对 BookList 进行增加操作,如果返回真,则执行第 42～45 行,把输入框设置为空,并且在界面上弹出一个对话框,第 45 行通过自定义的 builderDialogue 建立对话框。如果返回假,就生成一个消息对话框显示 "已有此图书"。第 52～65 行是建立对话框的函数实现,首先通过 Builder 类构成一个对话框,第 54 行设置对话框的标题,第 56～62 行设定对话框上按钮的事件处理,setNegativeButton 方法的第一个参数设定按钮的名称,第二个参数实现按钮的监听,当单击 "返回首页" 按钮时执行第 59 行的 finish()方法将当前 Activity 移出 Activity 栈。第 63 行定义取消按钮的名称,第 64 行调用 bulider 的 show()方法显示对话框。这里已经对 InsertActivity 界面做了详细的讲解。

下面讲解一下用 BookList 来模拟的存储数据的类,代码如下所示:

```
1.  package model.cqupt;
2.
3.  import java.util.ArrayList;
4.
5.  public class BookList extends ArrayList<Book> {
6.
7.      private static BookList booklist = null;
8.
9.      private BookList() {
10.         Book b1 = new Book("001", "c++", "24");
11.         Book b2 = new Book("002", "java", "45");
12.         Book b3 = new Book("003", "计算机", "21");
13.         add(b1);
14.         add(b2);
15.         add(b3);
16.     }
17.
18.     public static BookList getBookList() {
19.         if (booklist == null)
20.             booklist = new BookList();
21.         return booklist;
22.
23.     }
24. }
```

BookList 类采用了单例模式。单例模式也叫单子模式,是一种常用的软件设计模式。在应用这个模式时,单例对象的类必须保证只有一个实例存在。许多时候整个系统只需要拥有一个全局对象,这样有利于协调系统整体的行为。Chapter03 中只需要一个 booklist 对象用于存取数据,所以采用了单例模式。BookList 类继承了 ArrayList,并且声明一个私有的静态对象作为整个程序的全局对象。在第 18~23 行是一个静态的成员方法,外部可以直接通过类名调用,这个函数的功能是得到 booklist 对象,如果对象没有被创建过,那么就调用构造函数初始化,然后再返回 booklist 对象,如果已经创建了,就会直接返回 booklist 对象。booklist 对象在第 9~16 行的私有构造函数中初始化,私有的目的是防止外部调用构造函数,由于 BookList 继承了 ArrayList,所以可以直接调用 ArrayList 的 add 方法存入 3 本图书,这里将图书的信息封装到了 model.cqupt 包下的 Book 类中,Book 类代码如下所示:

```
1.  package model.cqupt;
2.
3.  public class Book {
4.
5.      private String name;
6.      private String id;
7.      private String price;
8.
9.      public Book(String id, String name, String price) {
10.         this.id = id;
11.         this.name = name;
12.         this.price = price;
13.     }
14.
15.     public String getName() {
16.         return name;
17.     }
18.
19.     public void setName(String name) {
20.         this.name = name;
21.     }
22.
23.     public String getId() {
24.         return id;
25.     }
26.
27.     public void setId(String id) {
28.         this.id = id;
29.     }
30.
31.     public String getPrice() {
32.         return price;
33.     }
```

```
34.
35.    public void setPrice(String price) {
36.        this.price = price;
37.    }
38.
39.    public String toString() {
40.        return id + " " + name + " " + price;
41.    };
42.
43. }
```

在 Book 类中有 set 和 get 函数分别设置和得到图书的信息。

Activity 界面类通过对 Controller 类的操控来对 BookList 进行操作。下面详细讲解 Controller 类。Controller 类的代码如下所示:

```
1.  package control.cqupt;
2.
3.  import model.cqupt.Book;
4.  import model.cqupt.BookList;
5.
6.  public class Controller {
7.      public boolean addBook(String id,String name,String price) {
8.          BookList bookList = BookList.getBookList();
9.          int i = 0;
10.         for (; i < bookList.size(); ++i) {
11.             Book book2 = bookList.get(i);
12.             String bid = book2.getId();
13.             if (bid.equals(id)) {
14.                 break;
15.             }
16.         }
17.         if (i == bookList.size()) {
18.             Book book = new Book(id, name, price);
19.             bookList.add(book);
20.             return true;
21.
22.         }
23.         return false;
24.     }
25.
26.     public BookList searchBook() {
27.         BookList bookList = BookList.getBookList();
28.         return bookList;
29.     }
```

```
30.
31.    public boolean deleteBook(String name)
32.    {
33.        BookList bookList = BookList.getBookList();
34.        for (int i=0; i<bookList.size(); ++i)
35.        {
36.            Book book2 = bookList.get(i);
37.            if(book2.getName().equals(name))
38.            {
39.                bookList.remove(i);
40.                return true;
41.            }
42.        }
43.        return false;
44.    }
45.    public boolean setBook(String id,String name,String price)
46.    {
47.        BookList bookList = BookList.getBookList();
48.        for(int i=0;i<bookList.size();++i)
49.        {
50.            Book book2 = bookList.get(i);
51.            if(book2.getId().equals(id))
52.            {
53.                Book book = new Book(id, name, price);
54.                bookList.set(i,book);
55.                return true;
56.            }
57.        }
58.        return false;
59.
60.    }
61. }
```

界面类通过 Controller 类操作 BookList。前面已经讲到了视图类中的 InsertActivity 和模型类中的 Booklist，这里将详细讲述控制类 Controller。Controller 在 Controller 中有 4 个方法，分别对 BookList 进行增、删、改、查操作。

增加图书（addBook），代码中第 7～24 行自定义了增加图书的方法，首先获得 booklist 对象，在第 9～23 行对 booklist 对象中图书进行循环遍历，如果在 booklist 对象中有与新增图书编号一样的图书，则跳出循环，并且返回 false，代表新增图书失败；如果 booklist 没有相同编号的图书，则执行第 17～22 行，对 booklist 操作增加图书，并且返回 true。

删除图书（deleteBook），代码第 31～34 行自定义了删除图书方法，与增加图书相同，首先获得 booklist 对象，在第 33～42 行循环遍历 booklist 对象中的图书名称，查看是否有需要删除的图书名称，如果有就返回 true，否则返回 false。

修改图书（setBook），代码中第 45～60 行自定义了修改图书的方法，首先获得 booklist 对象，然后在第 48～57 行循环遍历 booklist 对象中图书的编号，如果有此编号，那么就执行第 53～55 行对 booklist 对象进行修改操作，然后返回 true。如果没有这个编号的图书，则返回 false。

查询图书（searchBook），代码第 26～29 行自定义了查询图书的方法，与以上方法不同的是，查询图书只需要获得 booklist 对象，然后将此对象返回给 SearchActivity 界面，在 SearchActivity 界面中显示 booklist 对象中的图书，关于 SearchActivity 将在后面进行详细讲述。

下面将对剩下的 3 个视图类 DeleteActivity、SetActivity、SelectActivity 进行讲解。

首先讲解 DeleteActivity，界面如图 3-50 所示。

图 3-50　DeleteActivity

代码如下所示：

```
1.  package ui.cqupt;
2.
3.  import control.cqupt.Controller;
4.  import android.app.Activity;
5.  import android.app.AlertDialog.Builder;
6.  import android.content.DialogInterface;
7.  import android.os.Bundle;
8.  import android.view.View;
9.  import android.view.View.OnClickListener;
10. import android.widget.Button;
11. import android.widget.EditText;
12.
13. public class DeleteActivity extends Activity {
```

```
14.    private EditText name;
15.
16.    public void onCreate(Bundle savedInstanceState) {
17.        super.onCreate(savedInstanceState);
18.        setContentView(R.layout.delete);
19.        name = (EditText) findViewById(R.id.dname);
20.        Button delete = (Button) findViewById(R.id.d_delete);
21.        delete.setOnClickListener(new ButtonListener());
22.    }
23.
24.    class ButtonListener implements OnClickListener {
25.
26.        public void onClick(View v) {
27.            String bookname = name.getText().toString();
28.            Controller control = new Controller();
29.            if(bookname.equals(""))
30.            {
31.                new Builder(DeleteActivity.this).setMessage("图书名不能为空").show();
32.            }
33.            else
34.            {
35.                if (control.deleteBook(bookname)) {
36.                    name.setText("");
37.                    buildDialog();
38.                } else {
39.                    new Builder(DeleteActivity.this).setMessage("没有此图书").show();
40.                }
41.            }
42.        }
43.
44.        private void buildDialog() {
45.            Builder builder = new Builder(DeleteActivity.this);
46.            builder.setTitle("删除成功,是否继续删除图书");
47.            builder.setNegativeButton("返回首页",
48.                    new DialogInterface.OnClickListener() {
49.                        public void onClick(DialogInterface dialog,
50.                                int whichButton) {
51.                            finish();
52.                        }
53.
54.                    });
55.            builder.setPositiveButton("继续删除", null);
```

```
56.            builder.show();
57.        }
58.    }
59. }
```

DeleteActivity 继承了 Activity 类，并且覆写了 onCreate 方法初始化界面，通过 setContentView 方法设置了界面的布局，界面布局文件是 delete.xml 文件，代码如下所示：

```
1.  <?xml version="1.0" encoding="utf-8"?>
2.  <LinearLayout xmlns:android="http://schemas.android.com/apk/res/android"
3.      xmlns:tools="http://schemas.android.com/tools"
4.      android:layout_width="fill_parent"
5.      android:layout_height="fill_parent"
6.      android:gravity="center|top"
7.      android:orientation="vertical"
8.      tools:ignore="HardcodedText" >
9.
10.     <TextView
11.         android:layout_width="fill_parent"
12.         android:layout_height="wrap_content"
13.         android:text="输入删除的书名"
14.         android:gravity="center_horizontal"
15.         android:textSize="15pt" />
16.
17.     <TextView
18.         android:layout_width="wrap_content"
19.         android:layout_height="wrap_content"
20.         android:text="图书名称"/>
21.
22.     <EditText
23.         android:id="@+id/dname"
24.         android:layout_width="180dp"
25.         android:layout_height="wrap_content"
26.         android:background="#FFFFFF"
27.         android:ems="10"
28.         android:inputType="text"
29.         android:textColor="#000000" />
30.
31.     <Button
32.         android:id="@+id/d_delete"
33.         android:layout_width="102dp"
34.         android:layout_height="wrap_content"
35.         android:text="删除" />
```

36.
37. </LinearLayout>

在 delete.xml 文件中声明了 TextView、EditText 和 Button 组件，整个代码的风格与前面所讲的 insert.xml 文件相同，此处不再赘述。

现在回到之前的 DeleteActivity 代码上，onCreate 方法对在 delete.xml 文件中声明的组件实例化，并且在第 21 行为 Button 组件加上事件监听。第 24～58 行定义了一个 ButtonListener 的内部类来实现监听器，并且实现了 public void onClick(View v)方法进行事件处理。onClick 方法的实现和 InsertActivity 中的实现基本相同，此处不再赘述。

接下来是 SetActivity，界面如图 3-51 所示。

图 3-51 SetActivity

可以看到 SetActivity 的界面和 InsertActivity 的界面基本上一样（除了显示的文字不同），由此可知，SetActivity 的实现和 InsertActivity 的实现基本上相同。SetActivity 代码如下所示：

```
1.  package ui.cqupt;
2.
3.  import control.cqupt.Controller;
4.  import android.app.Activity;
5.  import android.app.AlertDialog.Builder;
6.  import android.content.DialogInterface;
7.  import android.os.Bundle;
8.  import android.view.View;
9.  import android.view.View.OnClickListener;
10. import android.widget.Button;
```

```
11.    import android.widget.EditText;
12.
13.    public class SetActivity extends Activity {
14.        private EditText name;
15.        private EditText id;
16.        private EditText price;
17.
18.        public void onCreate(Bundle savedInstanceState) {
19.            super.onCreate(savedInstanceState);
20.            setContentView(R.layout.set);
21.            name = (EditText) findViewById(R.id.sname);
22.            id = (EditText) findViewById(R.id.sid);
23.            price = (EditText) findViewById(R.id.sprice);
24.            Button set = (Button) findViewById(R.id.s_set);
25.            set.setOnClickListener(new ButtonListener());
26.        }
27.
28.        class ButtonListener implements OnClickListener {
29.
30.            public void onClick(View v) {
31.                String bookname = name.getText().toString();
32.                String bookid = id.getText().toString();
33.                String bookprice = price.getText().toString();
34.                Controller control = new Controller();
35.                if (bookname.equals("") || bookid.equals("")
36.                        || bookprice.equals("")) {
37.                    new Builder(SetActivity.this).setMessage("图书信息不
                        能为空").show();
38.                } else {
39.                    if (control.setBook(bookid, bookname, bookprice)) {
40.                        name.setText("");
41.                        id.setText("");
42.                        price.setText("");
43.                        buildDialog();
44.                    } else {
45.                        new Builder(SetActivity.this).setMessage("没有此编号的图
                            书,请重新输入")
46.                                .show();
47.
48.                    }
49.                }
50.            }
51.
52.            private void buildDialog() {
```

```
53.            Builder builder = new Builder(SetActivity.this);
54.            builder.setTitle("修改成功,是否继续修改图书");
55.            builder.setNegativeButton("返回首页",
56.                    new DialogInterface.OnClickListener() {
57.                        public void onClick(DialogInterface dialog,
58.                                int whichButton) {
59.                            finish();
60.                        }
61.
62.                    });
63.            builder.setPositiveButton("继续修改", null);
64.            builder.show();
65.        }
66.    }
67.
68. }
```

可以对比一下 InsertActivity 的代码，两者不同的是各个组件的名称和通过 control 对象对 booklist 进行的操作不同，其他功能完全一样。SetActivity 的布局文件是 set.xml 文件，代码如下所示：

```
1.  <?xml version="1.0" encoding="utf-8"?>
2.  <LinearLayout xmlns:android="http://schemas.android.com/apk/res/android"
3.      xmlns:tools="http://schemas.android.com/tools"
4.      android:layout_width="fill_parent"
5.      android:layout_height="fill_parent"
6.      android:gravity="center|top"
7.      android:orientation="vertical"
8.      tools:ignore="HardcodedText" >
9.
10.     <TextView
11.         android:layout_width="fill_parent"
12.         android:layout_height="wrap_content"
13.         android:gravity="center_horizontal"
14.         android:text="请输入图书信息"
15.         android:textSize="15pt" />
16.
17.     <TextView
18.         android:layout_width="wrap_content"
19.         android:layout_height="wrap_content"
20.         android:text="图书编号" />
21.
22.     <EditText
23.         android:id="@+id/sid"
```

```
24.        android:layout_width="180dp"
25.        android:layout_height="wrap_content"
26.        android:background="#FFFFFF"
27.        android:ems="10"
28.        android:inputType="text"
29.        android:textColor="#000000" />
30.
31.    <TextView
32.        android:layout_width="wrap_content"
33.        android:layout_height="wrap_content"
34.        android:text="修改名称" />
35.
36.    <EditText
37.        android:id="@+id/sname"
38.        android:layout_width="180dp"
39.        android:layout_height="wrap_content"
40.        android:background="#FFFFFF"
41.        android:ems="10"
42.        android:inputType="text"
43.        android:textColor="#000000" />
44.
45.    <TextView
46.        android:layout_width="wrap_content"
47.        android:layout_height="wrap_content"
48.        android:text="修改价格" />
49.
50.    <EditText
51.        android:id="@+id/sprice"
52.        android:layout_width="180dp"
53.        android:layout_height="wrap_content"
54.        android:background="#FFFFFF"
55.        android:ems="10"
56.        android:inputType="text"
57.        android:textColor="#000000" />
58.
59.    <Button
60.        android:id="@+id/s_set"
61.        android:layout_width="102dp"
62.        android:layout_height="wrap_content"
63.        android:text="修改" />
64.
65. </LinearLayout>
```

set.xml 和 insert.xml 基本相同，前面已经对 insert.xml 进行了详细讲解，这里就不再重复了，如果对 set.xml 代码还不是很清楚，那么请回顾一下对 insert.xml 的讲解。

接下来要讲解最后一个界面 SelectActivity。这个界面和 InsertActivity、DeleteActivity 有点不同，我们将会做详细的讲述。下面先看一下 SelectActivity 的界面，如图 3-52 所示。

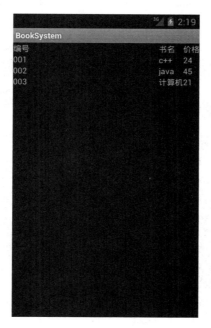

图 3-52　SelectActivity

SelectActivity 的代码如下所示：

```
1.  package ui.cqupt;
2.
3.  import control.cqupt.Controller;
4.
5.  import model.cqupt.Book;
6.  import model.cqupt.BookList;
7.
8.  import ui.cqupt.R;
9.
10. import android.app.Activity;
11. import android.os.Bundle;
12. import android.widget.TableLayout;
13. import android.widget.TableRow;
14. import android.widget.TextView;
15.
16. public class SelectActivity extends Activity {
17.
```

```
18.     public void onCreate(Bundle savedInstanceState) {
19.         super.onCreate(savedInstanceState);
20.         setContentView(R.layout.select);
21.         Controller control = new Controller();
22.         BookList booklist = control.searchBook();
23.         CreateTable(booklist);
24.     }
25.
26.     private void CreateTable(BookList booklist) {
27.       TableLayout table = (TableLayout) findViewById(R.id.SELECT_
          ACTIVITY_TableLayout);
28.         for (int i = 0; i < booklist.size(); ++i) {
29.             Book book = booklist.get(i);
30.             String id = book.getId();
31.             String name = book.getName();
32.             String price = book.getPrice();
33.             TableRow row = new TableRow(this);
34.             TextView tid = new TextView(this);
35.             TextView tname = new TextView(this);
36.             TextView tprice = new TextView(this);
37.             tid.setText(id);
38.             tname.setText(name);
39.             tprice.setText(price);
40.             row.addView(tid);
41.             row.addView(tname);
42.             row.addView(tprice);
43.             table.addView(row);
44.         }
45.     }
46. }
```

SelectActivity 和之前的界面不同点是在布局上，SelectActivity 采用了线性布局和表格布局，并且通过表格的形式显示图书。下面是 select.xml 文件代码：

```
1.  <?xml version="1.0" encoding="utf-8"?>
2.  <LinearLayout xmlns:android="http://schemas.android.com/apk/res/android"
3.      xmlns:tools="http://schemas.android.com/tools"
4.      android:layout_width="fill_parent"
5.      android:layout_height="fill_parent"
6.      android:orientation="vertical"
7.      tools:ignore="HardcodedText" >
8.
9.      <TableLayout
10.         android:id="@+id/SELECT_ACTIVITY_TableLayout"
11.         android:layout_width="fill_parent"
12.         android:layout_height="wrap_content"
```

```
13.            android:stretchColumns="0" >
14.
15.        <TableRow>
16.
17.            <TextView
18.                android:id="@+id/TextView01"
19.                android:layout_width="fill_parent"
20.                android:layout_height="wrap_content"
21.                android:text="编号" />
22.
23.            <TextView
24.                android:id="@+id/TextView02"
25.                android:layout_width="fill_parent"
26.                android:layout_height="wrap_content"
27.                android:text="书名" />
28.
29.            <TextView
30.                android:id="@+id/TextView03"
31.                android:layout_width="fill_parent"
32.                android:layout_height="wrap_content"
33.                android:text="价格" />
34.        </TableRow>
35.    </TableLayout>
36. </LinearLayout>
```

可以看到，上面代码的第 9～35 行是将所有的组件都放入<TableLayout></TableLayout>中，这是一个表格布局。表格布局中<TableRow></TableRow>标签，这个标签代表了表格的一行，每一行的内容都要在 TableRow 中显示，SelectActivity 的显示功能采用了动态增加组件的方式。SelectActivity 继承了 Activity 类并且覆写了 onCreate 方法，在第 22 行通过 control 的 searchBook 方法得到 booklist 对象的引用，将 booklist 对象作为参数传递给自定义创建表格的方法 CreateTable。第 26～46 行是 CreateTable 方法的实现，创建表格显示 booklist 对象中的图书。首先实例化 TableLayout，然后在第 28～44 行循环遍历 booklist，为每个 booklist 中的 book 创建一个 TableRow，代表表格的一行内容，第 30～32 行提取 booklist 中图书信息，在第 34～36 行为图书的编号、名称和价格创建显示的组件 TextView，然后把图书信息显示到 TextView 组件中，在第 40～42 行把显示组件 TextView 加入到代表表格中每行的 TableRow 中，最后将这一行加入到表格布局中，完成对图书信息的显示功能。

这里完成了对 Chapter03 的讲解，在以后的学习中，我们会在 Chapter03 的基础上进行修改，加入学习的新知识。

第 4 章　数 据 存 储

数据存储是应用程序最基本的问题,在开发过程中应用最频繁,我们知道任何企业系统、应用程序都不可避免地要用到大量的数据,所以数据存储必须以某种方式保存不能丢失,并且能够有效、简便地使用和更新这些数据。Android 提供了如下 4 种数据存储的方式:Peference(配置)、File(文件)、SQLite 数据库和 Network(网络)。由于存储的数据在应用程序中都是私有的,并且各个应用程序都是相互独立的,所以各个应用程序之间的数据是不能实现数据共享的,如果实现不同应用程序之间的数据共享就必须使用 Android 提供的 Content Provider 组件。本章主要讲解数据存储的前 3 种方式以及不同应用程序之间的数据如何共享。

4.1　Preference 存储方式

在应用程序的使用过程中,用户会经常根据自己的习惯爱好更改应用程序的设置,为了能够保存用户的设置内容,应用程序一般都会在文件系统中保存一个配置文件,并且每次在应用程序启动的时候读取这个配置文件的内容。常见的操作系统应用软件中,用于保存配置的文件类型一般是 INI(INI 是配置文件所采用的存储格式,用来配置应用软件以实现不同用户的要求。一般不用直接编辑这些 INI 文件,它可以用来存放软件信息,注册表信息等)或者 XML[XML 即可扩展标记语言 (Extensible Markup Language,XML),用于标记电子文件使其具有结构性的标记语言,可以用来标记数据、定义数据类型,是一种允许用户对自己的标记语言进行定义的源语言]的文件格式,当然我们也可以自己定义,INI 格式简单易懂但需要写代码实现文件的写入和读取,XML 在代码上比较容易实现,但是可读性不如 INI。尽管它们都有各自的优点,但是它们都需要程序员进行复杂的代码实现。

所以 Android 为开发人员提供了更为精简的数据存储方式 Preference(配置)。下面来看看什么是 Preference 以及它的应用。

Android 中的 Preference 提供了一种轻量级(轻量级:一般是指规模较小,或者功能较为完善,开启程序时所需要的资源比较小)的数据存取方法,应用场合则是数据比较少的系统配置信息,它的本质是基于 XML 文件存储 Key-Value(键值对)数据的存储方式。它只能存储基本的数据类型。特别注意:它无法直接在多个程序间共享 Preference 数据。

4.1.1 SharedPreferences

1. SharedPreferences 是什么

SharedPreferences 是 Android 平台上的一个轻量级的存储类，主要是保存一些界面配置信息，例如窗口状态，一般在 Activity 中重载窗口状态 onSaveInstanceState 保存，一般使用 SharedPreferences 完成。例如，当我们在一个用户资料的 Activity 界面中填写完自己的个人信息之后，退出程序时，填写的资料被保存，再次打开该界面的时候输入的资料就会显示出来。SharedPreferences 提供了 Android 平台常规的 Long（长整型）、（Int 整型）、String（字符串型）的保存。

2. 使用 SharedPreferences 存取数据

使用 Preference 方式来存取数据，其实就是处理一个 Key-Value。它主要用到了 SharedPreferences 接口和 SharedPreferences 的一个内部接口 SharedPreference.editor。SharedPreferences 接口提供了方法实现数据的获取，SharedPreference.editor 接口提供了方法实现数据的保存以及修改（注意：Preference 主要针对系统配置信息的保存，SharedPreferencse 对象本身只能获取数据而不支持存储和修改，存储批改是经由过程 Editor 对象实现）。使用 SharedPreferences 存储数据的步骤如下：

（1）调用 getsharedPreferences(String name,int mode)方法得到 SharedPreferences 的对象。其中第一个参数 String name 是 Key-Value 数据所保存位置文件的文件名，即 Key-Value 数据保存的文件名由第一个参数给出。第二个参数是操作模式，一共有 3 个操作模式：MODE_PRIVATE（私有模式）、MODE_WORLD_READABLE（全局读模式）以及 MODE_WOELD_WRITEABLE（全局写模式）。如果定义为私有则之后创建程序有权限进行读和写，如果定义为全局读，则除了创建程序可以读写之外其他程序也可以进行读操作，但不可以写操作；如果定义为全局写，则创建程序和其他程序都可以进行写操作，但不可以进行读操作。定义操作模式代码如下：

定义私有模式——MODE_PRIVATE；定义全局读模式——MODE_WORLD_READABLE；定义全局写模式——MODE_WOELD_WRITEABLE；定义为既可全局读又可全局写模式——MODE_WORLD_READABLE+ MODE_WOELD_WRITEABLE。

（2）利用 SharedPreferences 接口的 edit()方法获取 SharedPreferences.Editor 对象。

（3）通过 SharedPreferences.Editor 对象保存 Key-Value 数据。

（4）通过 SharedPreferences.Editor 接口的 commit()方法提交数据（注意：只有在调用了 commit()方法之后才会真正将 Key-Value 数据保存在相应的文件中）。

3. SharedPreferences 存储方式下数据的存储位置

前面已经讲过使用 SharedPreferences 存储数据，但是数据是存储在什么地方呢？这里就来说明数据存储的位置。对于软件配置参数的保存，如果是 Windows 软件，通常会采用 INI 文件进行保存。如果是 J2SE 应用，则会采用 Properties 属性文件进行保存。但是如果是 Android 应用，那么最适合采用什么方式保存软件配置参数呢？使用 SharedPreferences 保存数据，其背后是用 XML 文件存放的数据，文件存放在该应用程序下/data/data/<package name>/shared_prefs 目录下（特别注意：只有应用程序中使用了

Preference 才会在该文件夹下面产生一个 shared_prefs，其中就是我们保存的数据。查看其的方法为：右击 project 选择"运行"命令，在 Eclipse 中切换到 DDMS 视图，选择 FILe Explore 标签找到/data/data/包名下面的 shared_prefs 即可，此时的 Android 虚拟机是处于运行状态的），如图 4-1 所示。

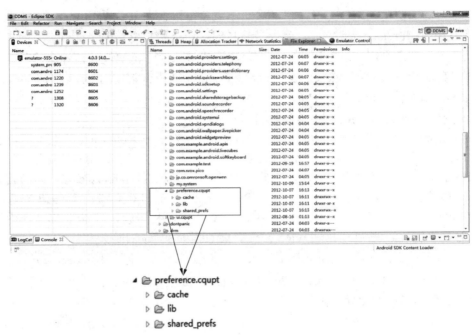

图 4-1　SharedPreferences 生成的数据文件存储目录

单击 按钮可以导出 XML 文件，如图 4-2 所示。

图 4-2　导出 XML 文件

4.1.2　PreferenceActivity

Android 系统内的设置界面由 Android Preference 的相关类提供,虽然使用 Preference 这个 Key-Value 的方式来自动保存这些数据，并即时生效，但是 Android 也提供了一种类似 Layout 的方式来进行 Preference 的布局，并且能够透明地保存信息。那就是 PreferenceActivity。其中，PreferenceActivity 是 Activity 的子类，该类封装了 SharedPreferences，因此 PreferenceActivity 的所有子类都会有保存 Key-Value 的能力。PreferenceActivity 提供了一些常用的设置项，设置子项包含以下种类：ListPreference、CheckBoxPreference、EditTextPreference 等。ListPreference 对应<ListPreference>标签；单击该设置项会弹出一个带 Listview 的对话框，CheckBoxPreference 对应<CheckBoxPreference>标签；单击该设置项会弹出一个带 CheckBox 的组件，EditTextPreference 对应<EditTextPreference>标签；单击该设置项会弹出一个带 EditText

组件的对话框。这些设置项可以满足大多数配置界面的要求。由此可见，Android Preference 向用户提供一些参数设置的接口，使用 Preference 相关的一些类，就可以很方便地呈现参数设置界面以及对参数的设置进行处理。

使用 PreferenceActivity 方式保存参数的步骤大致如下：

（1）在 XML 中配置参数界面元素。

（2）加载 XML 中的参数元素。

（3）在 Java 代码中使用相关的方法操作参数元素。

下面来看一个示例，如图 4-3～图 4-5 所示。

图 4-3　PreferenceActivity 设置界面

图 4-4　子界面

图 4-5　对话框

下面来看看这个示例是如何实现的。

首先要知道是如何创建 XML 参数配置文件的。该 XML 文件是通过在 Project 下的

res 文件夹下先新建一个目录文件夹 XML，然后选中该文件夹，右击，选择 New→Other→Android→Android XML File 选项，得到如图 4-6 所示的界面。

图 4-6　新建 Android XML 文件

☞**注意**：Resource Type（资源类型）选择 Preference，File（文件名）在此命名为 set_preference，在 Root Element 中保持默认，即选择 PreferenceScreen，否则在 Activity 中绑定该资源时，将报类似"java.lang.RuntimeException: Unable to start activity……"的错误。最后单击 Finish 按钮，完成新建 Android XML 配置文件的工作。添加完成之后，在 res/xml/下打开添加的 set_preference.xml 文件，可以看到 Android 也为我们提供了两种编辑模式：可视化的结构设计（structure）及 xml 源码设计。推荐使用 structure 进行创建。如图 4-7 所示。

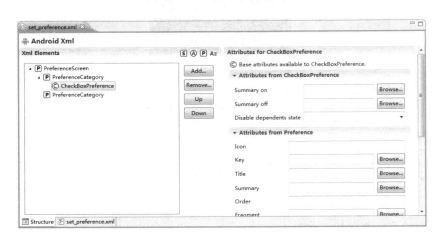

图 4-7　可视化编辑结构图

可以利用 Add 按钮添加各种配置元素，右边则是它的编辑属性。如图 4-7 所示。

无论是使用可视化编辑还是直接在 XML 文件中编辑，都可以得到 set_preference 的 XML 代码：

```xml
1.  <?xml version="1.0" encoding="utf-8"?>
2.  <PreferenceScreen xmlns:android="http://schemas.android.com/apk/res/android">
3.      <PreferenceCategory android:title="我的主页">
4.          <CheckBoxPreference android:title="使用个性签名"
5.                  android:summary="使用签名设计软件设计个性签名"
6.                  android:key="sign_name"/>
7.          <CheckBoxPreference android:title="启用流动网络设置"
8.                  android:summary="确定上网"
9.                  android:key="data_setting"/>
10.     </PreferenceCategory>
11.     <PreferenceCategory android:title="个人信息设置">
12.         <CheckBoxPreference android:title="是否保存个人信息"
13.                 android:key="yesno_save_individual_info"/>
14.         <EditTextPreference android:title="姓名"
15.                 android:key="individual_name"
16.                 android:summary="请输入真实姓名"/>
17.         <PreferenceScreen android:summary="是否在校学习、手机"
18.                 android:key="other_student_msg"
19.                 android:title="其他个人信息">
20.             <CheckBoxPreference android:title="是否在校学习"
21.                     android:key="is_a_student"/>
22.             <EditTextPreference android:title="手机"
23.                     android:key="mobile"
24.                     android:summary="请输入您的真实手机号码"/>
25.         </PreferenceScreen>
26.     </PreferenceCategory>
27. </PreferenceScreen>
```

PreferenceActivity 是专门用于显示 Preference 的，所以只要创建一个继承自 PreferenceActivity 类即可将 XML 文件中的参数元素加载进来。在 Java 中的代码如下：

```java
1.  package PreferenceActivity.a;
2.  
3.  import android.app.Activity;
4.  import android.os.Bundle;
5.  import android.preference.PreferenceActivity;
6.  
7.  public class PreferenceActivityActivity extends PreferenceActivity{
8.      @Override
9.      public void onCreate(Bundle savedInstanceState){
```

```
10.        super.onCreate(savedInstanceState);
11.        // 所创建的值将会自动保存到 SharePreferences
12.        addPreferencesFromResource(R.xml.set_preference);
13.    }
14. }
```

PreferenceActivityActivity 继承了 PreferenceActivity 类，在 onCreate()方法中不需要设置布局文件，只需要执行第 12 行的 addPreferencesFromResource 方法装载 XML 文件即可。

下面来看一看 PreferenceActivity 是怎样设置界面的。首先介绍参数界面元素，接着讲解如何在 XML 中配置参数元素和加载 XML 文件，以及怎样在 Java 代码中操作参数元素和 XML 文件的保存路径。

4.1.3 XML 解析

XML 现在已经成为一种通用的数据交换格式，它的平台无关性、语言无关性、系统无关性，给数据集成与交互带来了极大的方便。XML 在不同的语言中解析方式都是一样的，只不过实现的语法不同。本节介绍对 XML 的两种解析方式：DOM 解析器、SAX 解析器。

1. DOM 解析器

DOM 是基于树结构的节点或信息片段的集合，根据 DOM API 为 XML 文档的解析定义了一组接口来遍历 XML 树、检索所需数据。使用 DOM 解析 XML 通常需要将整个 XML 文档加载到内存，然后利用 DOM 中的对象，对 XML 文档进行读取、搜索、修改、添加和删除等操作。

DOM 的工作原理是：使用 DOM 对 XML 文件进行操作时，首先要解析文件，将文件分为独立的元素、属性和注释等，然后以节点树的形式在内存中对 XML 文件进行表示，就可以通过节点树访问文档的内容，并根据需要修改文档。

DOM 在内存中以树结构存放，因此检索和更新效率会更高。但是对于特别大的文档，解析和加载整个文档将会很耗费资源。当然，如果 XML 文件的内容比较小，采用 DOM 是可行的。

常用的 DOM 接口和类如下：

Document——该接口定义分析并创建 DOM 文档的一系列方法，它是文档树的根，是操作 DOM 的基础。

Element——该接口继承 Node 接口，提供了获取、修改 XML 元素名字和属性的方法。

Node——该接口提供处理并获取节点和子节点值的方法。

NodeList——提供获得节点个数和当前节点的方法。这样就可以迭代地访问各个节点。

DOMParser——该类是 Apache 的 Xerces 中的 DOM 解析器类,可直接解析 XML 文件。

User.xml 如下所示：

```
<user>
<name>Lee</name>
<age>18</age>
<sex>male</sex>
</user>
<user>
<name>vivian</name>
<age>22</age>
<sex>female</sex>
</user>
```

User.xml 的 DOM 解析流程图如图 4-8 所示。

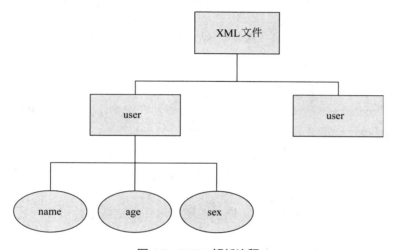

图 4-8　DOM 解析流程

2．SAX 解析器

上面提到 DOM 解析是将整个文档读入内存中，这样就保留了过多的不需要的节点，浪费了内存和空间。为了解决 DOM 解析存在的问题，就出现了 SAX 解析。SAX（Simple API for XML）解析器是一种基于事件模型的解析器，它将 XML 文档转化为一系列的事件，每个事件由单独的事件处理器来决定如何处理。例如，对文档进行顺序扫描，当扫描到文档开始与结束、元素开始与结束等地方时通知事件处理函数，由事件处理函数做相应动作，然后继续同样的扫描，直至文档结束。

SAX 解析器的优点是解析速度快、占用内存少，非常适合在 Android 移动设备中使用。

在 SAX 接口中，事件源是 org.xml.sax 包中的 XMLReader，它通过 parser()方法来解析 XML 文档，并产生事件。事件处理器是 org.xml.sax 包中 ContentHander、DTDHandler、ErrorHandler 以及 EntityResolver 这 4 个接口。XMLReader 通过相应事件处理器注册方法 setXXXX()来完成与 ContentHandler、DTDHandler、ErrorHandler 以及 EntityResolver 这 4 个接口的连接，如图 4-9 所示。

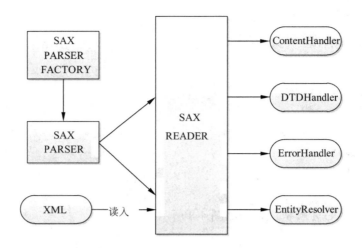

图 4-9 SAX 工作原理

常用的 SAX 接口和类如下：

Attributes——用于得到属性的个数、名字和值。

ContentHandler——定义与文档本身关联的事件（例如，开始和结束标记）。

DTDHandler——定义与 DTD 关联的事件。

DeclHandler——是 SAX 的扩展，但不是所有的语法分析器都支持它。

EntityResolver——定义与装入实体关联的事件，只有少数几个应用程序注册这些事件。

ErrorHandler——定义错误事件，许多应用程序注册这些事件以便用它们自己的方式报错。

DefaultHandler——提供了 SAX 接口的默认实现，详见表 4-1。DefaultHandler 覆盖相关的方法要比直接实现一个接口更容易。

表 4-1 DefaultHandler 方法

方　　法	说　　明
Setdocumentlocator(Locator locator)	设置一个可以定位文档内容事件发生位置的对象
StartDocument()	开始文档解析
StartElement(string uri, string localname, string qname, attributesatts)	处理元素开始事件，从参数中可以获得空间的 uri、元素名称列表等信息
Characters(char[]ch, int start, int length)	处理元素的字符内容，从参数中可以获取具体内容
EndelEment(string uri, string localname, string qname)	处理元素结束事件，从参数中可以获得所在空间的 uri、元素名称等信息
EndDocument()	用于处理文档解析的结束事件

可知，我们需要 XmlReader 以及 DefaultHandler 来配合解析 XML。

SAX 的解析流程如图 4-10 所示。

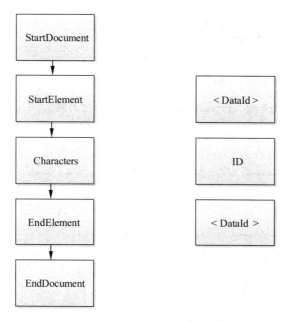

图 4-10　SAX 解析过程

4.2　文件的存储

前面已经讲过 Preference（配置）在 Android 中是一种简单的数据存储，存储的内容是一些 Key-Value 数据。Preference 简单键值存储形式，以 XML 格式存储在手机中，这是一个简单、方便、好操作的数据存储工具，但是只能存储简单的数据，如果存储大量数据就不方便了。这时可以采用文件存储，在 Android 系统中，可以很方便地利用文件存储我们想要的内容。Android 系统使用的是基于 Linux 系统的文件系统，程序员可以建立和访问自身私有文件的内部文件，也可以访问 SD 卡文件等外部文件。本节讲解文件的内部存储和外部存储。

4.2.1　内部存储

内部存储（Internal Storage）是指在物理上 Android 内置的闪存（不同于普通意义上的内存），外部存储一般指添加的 SD 卡（极少部分手机没有 SD 卡，但也有/sdcard 分区）。内部存储就是将文件保存在设备内部存储器中，但是这样的数据是存储在应用程序内的，也就是说，对这样存储的文件大小有一定的限制。默认情况下，这些文件是相应程序私有的，对其他程序不透明，对用户也是不透明的。当程序卸载后，这些文件就会被删除。内部存储与其他的（外部存储）相比有着比较稳定、存储方便、操作简单、更加安全（因为可以控制访问权限）等优点。

Context 中提供了相应的函数对文件进行操作，如表 4-2 所示。

表 4-2 Context 操作函数

方 法 名 称	描 述
FileInputStream openFileInput(String name)	打开应用程序的数据文件夹下的 name 文件对应的输入流
FileOutputStream openFileOutput(String name, int mode)	打开应用程序的数据文件夹下的 name 文件对应的输出流,并且指定以某种方式打开
getDir(String name, int mode)	在应用程序的数据文件下获取或创建 name 对应的子目录
File getFilesDir()	得到该应用程序数据文件夹的绝对路径
String[] fileList()	得到该应用程序数据文件夹下的全部文件
deleteFile(String name)	删除该应用程序的数据文件夹下的指定文件

在内部存储器中创建并保存数据文件,可以按照以下步骤来做:

(1)调用 openFileOutput()方法。

(2)使用 write()方法向文件中写入数据。

(3)调用 close()方法,关闭输出流。

从内部存储器中读取数据的步骤如下:

(1)调用 openFileInput()方法。

(2)调用 read()方法读取字节。

(3)调用 close()方法关闭输入流。

4.2.2 外部存储

所有 Android 设备都支持可以保存文件的共享外部存储(External Storage)。外部存储器可以是可移动存储器(如 SD 卡),也可以是内置在设备中的外部存储器(不可移动)。外部存储器上的文件是全部可读的,当设备通过 USB 连接计算机和计算机互传文件时,外部存储器上的文件是不可修改的。当外部存储器被挂载到计算机上或被移除,文件对 Android 设备就不可见了,且此时外部存储器上的文件是没有安全保障的。所有程序都可以读写外部存储器上的文件,用户也可以删除这些文件。与内部存储不同的是,当程序卸载时,它在外部存储所创建的文件数据是不会被清除的。

前面了解了 Android 的数据存储采用文件,但是这样的数据是存储在应用程序内的。也就是说,对这样存储的文件大小有一定的限制,有时候需要存储更大的文件。例如视频(电影)、音频文件等,这就用到了 SD 卡。Android 也提供了 SD 卡的一些相关操作,如表 4-3 和表 4-4 所示。

表 4-3 SD 卡的常用常量

常 用 常 量	说 明
String MEDIA_MOUNTED	当前 Android 的外部存储器可读可写
String MEDIA_MOUNTED_READ_ONLY	当前 Android 的外部存储器只读

表 4-4　SD 卡的常用方法

方 法 名 称	描　述
public static File getDataDirectory()	获得 Android 下的 data 文件夹的目录
public static File getDownloadCacheDirectory()	获得 Android Download/Cache 内容的目录
public static File getExternalStorageDirectory()	获得 Android 外部存储器也就是 SDCard 的目录
public static String getExternalStorageState()	获得 Android 外部存储器的当前状态
public static File getRootDirectory()	获得 Android 下的 root 文件夹的目录

要想实现对 SD 卡的读取操作，只需要按以下几个步骤进行：

（1）判断这台手机设备上是否有 SD 卡且具有读写 SD 卡的权限。通过调用 Environment 类的 getExternalStorageState()方法获取手机 SD 卡状态，如果手机中有 SD 卡，则返回的状态等价于 Environment 类的静态常量 MEDIA_MOUNTED。

（2）调用 Environment 类的 getExternalStorageDirectory()方法获取外部存储器的目录。

（3）使用 IO 流对外部存储器进行文件的读写。

（4）特别注意要在 AndroidMainfest.xml 中添加权限。

```
<uses-permission android:name="android.permission.WRITE_EXTERNAL_STORAGE"/>
```

4.3　SQLite 数据库

4.3.1　SQLite 简介

对于 SQLite，SQ 为 Structured Query（结构化查询）的缩写，Lite 表示轻量级。SQLite 是一款开源的关系数据库，是遵守 ACID[ACID 是数据库事务正确执行的 4 个基本要素的缩写，即原子性(Atomicity)、一致性（Consistency）、隔离性（Isolation）、持久性（Durability）]的关联式数据库管理系统，它的设计目标是嵌入式的，目前已经在很多嵌入式产品中使用，它占用的资源非常少，在嵌入式设备中，可能只需要几百千字节的内存就够了。它能够支持 Windows/Linux/UNIX 等主流的操作系统，同时能够与很多程序语言相结合，例如 Tcl、C#、PHP、Java 等，同样比起 MySQL、PostgreSQL 这两款世界著名的开源数据库管理系统来讲，它的处理速度更快，效率更高。SQLite 能够对大量数据进行存储，方便操作，SQLite 是 Android 内置的一个很小的关系数据库。

下面是访问 SQLite 的官方网站 http://www.sqlite.org/时第一眼看到的关于 SQLite 的特性：

（1）ACID 事务。

（2）零配置，即无须安装和管理配置。

（3）存储在单一磁盘文件中的一个完整的数据库。

（4）数据库文件可以在不同字节顺序的机器间自由地共享。

（5）支持数据库大小至 2TB。

（6）足够小，大致 3 万行 C 代码，250KB，运行时占用内存大概几百千字节。

（7）比一些流行的数据库大部分操作要快。
（8）简单方便的 API。
（9）包含 TCL 绑定同时通过 Wrapper 支持其他语言的绑定。
（10）良好注释的源代码并且有着 90%以上的测试覆盖率。
（11）独立没有额外依赖。
（12）完全开源，可以用于任何用途，包括出售它。
（13）支持多种开发语言，如 C 语言、PHP、Java、ASP.NET。

SQLite 数据库的优势在于它可嵌入到其他应用程序中，这样不仅提高了运行效率，而且屏蔽了数据库使用和管理的复杂性，程序只需要运行简单的基本的数据操作，其他操作交给系统内部数据库处理。注意：SQLite 数据库能存储较多的数据，能将数据库文件存放到 SD 卡中。

SQLite 数据库结构体系组成如图 4-11 所示，图中显示了 SQLite 的主要组成部件及其相互关系。下面将描述每一个部件。

总体来看，SQLite 分为 3 个子系统，其中包含了 8 个独立模块：接口（Interface）、词法分析器（Tokenizer）、语法分析器（Parser）、代码生成器（Code Generator）、虚拟机（Virtual Machine）、B 树（B Tree）、页缓存（Pager）、操作系统接口（OS Interface）图中 Complier 为编译器，Core 为核心模板，Backend 为后端。

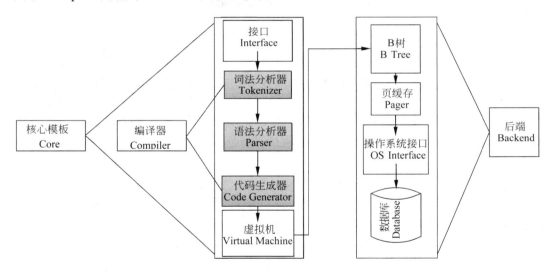

图 4-11 SQLite 数据库体系结构

1．接口

虽然有些函数分布在其他的文件中，但是主要的 SQLite 库的公用接口函数是在 main.c、legacy.c 和 vdbeapi.c 源代码文件中实现的。sqlite3_get_table()函数在 table.c 中实现，sqlite3_mprintf()在 printf.c 中实现，sqlite3_complete()是在 tokenize.c 中实现的。Tcl 接口在 tclsqlite.c 中实现。关于 SQLite 的接口更完整的信息在 http://www.sqlite.org/capi3ref.html 中有描述。为避免和其他软件的命名冲突，SQLite 库中所有的外部符号都以 sqlite3 为前缀，这些符号的目的就是为外部使用。换句话说，所有以 sqlite3_开始的符号，形成了 SQLite 的 API。

2．词法分析器

当一个 SQL 语句执行时，接口首先把包含该 SQL 语句的字符串传给词法分析器来进行处理。词法分析器负责把字符串分成一个一个的词法单元，然后把词法单元传递给语法分析器，词法分析器是在 tokenize.c 文件中实现的，这些代码是人工编写的，而不是使用 lex 之类的工具生成的。需要注意的是，在本设计中，词法分析器调用语法分析器，熟悉 YACC 和 BISON 的人一般总是在语法分析器中调用词法分析器。SQLite 的作者试过这两种方法，发现在词法分析器中调用语法分析器更好。

3．语法分析器

语法分析器根据上下文对词法分析出来的单元理解其含义。SQLite 的语法分析器是使用 Lemon(http://www.hwaci.com/sw/lemon/)的 LALR(1)工具产生器生成的。Lemon 和 YACC/BISON 工具差不多，但是 Lemon 使用一种不同的输入语法，这种语法更难以出错。Lemon 能产生一个可重入和线程安全的语法分析器，Lemon 定义了一种非终结符析构器，以致在语法出现错误时不至于出现内存泄漏。Lemon 分析器的输入文件在 parse.y 中定义。由于 Lemon 不是一个常见的程序，其完整的源代码仅为一个 C 文件，在 SQLite 的 tool 子目录中。Lemon 的文档在 doc 子目录中。

4．代码生成器

在语法分析器分析完 SQL 语句后，它调用代码生成器来生成在虚拟机上执行的代码，这些代码是按照 SQL 语句的要求来执行的。代码生成器包含在许多文件中：attach.c、auth.c、build.c、delete.c、expr.c、insert.c、pragma.c、select.c、trigger.c、update.c、vacuum.c 和 where.c。expr.c 处理表达式的代码生成，where.c 处理 SELECT、UPDATE 和 DELETE 语句中的 WHERE 子句的代码生成，attach.c、delete.c、insert.c、select.c、trigger.c、update.c 和 vacuum.c 处理与其名字相同的 SQL 语句的代码生成，这其中的每个文件在必要时都调用 expr.c 和 where.c 中的函数。其他的 SQL 语句在 build.c 中实现，auth.c 文件实现 sqlite3_set_authorizer()函数的功能。

5．虚拟机

代码生成器产生的程序在虚拟机上运行，该虚拟机的信息在文档 http://www.sqlite.org/opcode.html 中描述。概括来讲，虚拟机实现了一个抽象的计算引擎，这个计算引擎用来操作数据库文件。虚拟机有一个栈用于保存计算的中间状态，每条指令包括一个操作码和最多三个操作数。虚拟机在 vdbe.c 中实现。虚拟机有它自己的头文件：vdbe.h 文件定义了虚拟机和 SQLite 库的接口，vdbeInt.h 文件定义了虚拟机的结构。vdbeaux.c 文件中包含一些虚拟机和接口模块使用的工具。vdbeapi.c 文件包含了虚拟机的外部接口，例如 sqlite3_bind_...之类的函数。字符串，整数，浮点数，BLOB 类型都被存储在一个名为 Mem 的内部对象中，这个内部对象在 vdbemem.c 文件中实现。SQLite 使用回调 C 语言函数的方法来实现 SQL 语句的功能。甚至内建的 SQL 功能也是通过这种方法来实现。大部分 SQL 内建的函数，例如 coalesce()、count()、substr()等，在 func.c 中实现。日期和时间转换函数在 date.c 中实现。

6．B 树

SQLite 使用 B 树来实现数据库，B 树在 btree.c 文件中实现。数据库中的每个表和索引都使用一个单独的 B 树。所有的 B 树都存放在一个磁盘文件中。该数据库文件格式的细节在 btree.c 文件开始部分的注释里详细描述。B 树子系统的接口在头文件 btree.h 中定义。

7．页缓存

B 树模块使用固定的块大小从磁盘中请求信息。默认块大小为 1024B，但是可以从 512～65 536B 调整。页缓存负责读、写和缓存这些块。页缓存也提供了回滚和原子提交的功能抽象和数据库文件的锁操作。B 树驱动程序从页缓存中取得页，并且通知页缓存程序何时修改，提交或回滚操作，页缓存处理所有的这些麻烦细节，确保请求被快速、安全和高效地处理。

实现页缓存机制的代码在单个 C 文件 pager.c 中。页缓存子系统的接口在 pager.h 中定义。

8．操作系统接口

为了提高在 POSIX 和 Win32 系统中的可移植性，SQLite 和操作系统的接口使用了一个抽象层。此抽象层的接口在 os.h 中定义。每个操作系统有其自己的实现：os_unix.c 是 UNIX 系统的，os_win.c 是 Windows 系统的等。每个操作系统相关的实现都有其自己的头文件，如 os_unix.h、os_win.h 等。

以上就是 SQLite 数据库的简单介绍，对 SQLite 数据库的结构体系大致了解一下即可，不必深入研究。

4.3.2　SQLite 数据库基本数据操作

SQLite 数据库的操作一般包括创建、打开、关闭、删除数据库以及对 SQLite 数据库中表的一些方法的操作。

1．创建和打开数据库

创建和打开数据库，调用 SQLiteDatabase 类的 openOrCreateDatabase（String name，int mode，SQLiteDatabase.CursorFactory factory）方法，name 为数据库的名字，mode 为权限。如果数据库不存在，则会创建新数据库；如果存在，则打开数据库。

2．关闭数据库

对数据库操作完毕之后，就要关闭数据库，否则会抛出 SQLiteException 异常。关闭数据库只需调用 SQLiteDatabase 类的 close()方法即可。

3．删除数据库

删除数据库调用 SQLiteDatabase 类的 deleteDatebase()方法即可。

4．SQLite 数据库中表（Table）的操作

首先要明确的一点是，一个数据库可以有很多表，一个表中包含很多条数据。也就是说，在数据库中数据其实是保存在表中的。对数据库表的操作一般包括：创建表，向表中添加数据，从表中删除数据，修改表中的数据，查询表中的数据。

1）创建一个表

通过调用数据库的 execSQL(String sql)方法可以创建一个表，关于 execSQL(String sql)

方法的详细可以参看以下说明：

```
public void execSQL(String sql)
```

事实上，execSQL(String sql)方法的参数 sql 是 SQL 语句，为字符串类型。例如：

```
1.  SQLiteDatabase db;
2.  String sql = "CREATE TABLE book(" + "id VARCHAR(30) NOT NULL, "
    i.+ "name VARCHAR(30) NOT NULL, "
    ii.+ "price VARCHAR(30) NOT NULL) ;"
3.  db.execSQL(sql);
```

2）向表中插入数据

向数据库表中插入数据，可以直接调用 SQLiteDatabase 类的 insert()方法，也可以通过 execSQL(String sql)执行插入操作的 SQL 语句。

```
public long insert(String table, String nullColumnHack, ContentValues values);
```

table 是表名，第二个参数一般为 null，第三个参数是 ContentValues。若成功插入，则返回新插入行的 id，否则返回-1。例如：

```
1.  ContentValues cv=new contentValues();
2.  cv.put(num,1);
3.  cv.put(data,"测试一下数据库");
4.  db.insert(table,null,cv);
```

3）删除表中的一条数据

可以直接调用 SQLiteDatabase 类的 delete()方法，也可以通过 execSQL(String sql)执行删除操作的 SQL 语句。

```
public intdelete(String table, String whereClause, String[] whereArgs)
```

第一个参数是表名。第二个参数是删除的条件，如果为 null，则所有行都将删除。第三个参数是字符串数组，与 whereClause 配合使用。

4）修改表中数据

可以调用 SQLiteDatabase 类的 update()方法，也可以通过 execSQL(String sql)执行修改操作的 SQL 语句。

```
public int update(String table,ContentValues values,String whereClause,
String[] whereArgs)
```

5）查询数据

调用 SQLiteDatabase 类的 query()方法。

```
public Cursor query(String table,String[]columns,String selection,
String[]selectionArgs,String groupBy,String having,String orderBy,
String limit);
```

参数说明：

table——要查询数据的表名。

columns——要返回列的列名数组。

selection——可选的 where 子句，如果为 null，将会返回所有的行。

selectionArgs——当在 selection 中包含 "?" 时，如果 selectionArgs 的值不为 null，则这个数组中的值将依次替换 selection 中出现的 "?"。

groupBy——可选的 group by 子句，如果其值为 null，将不会对行进行分组。

having——可选的 having 子句，如果其值为 null，将会包含所有的分组。

orderBy——可选的 order by 子句，如果其值为 null，将会使用默认的排序规则。

limit——可选的 limit 子句，如果其值为 null，将不会包含 limit 子句。

```
Cursor cr=db.query("pic", null, null, null, null, null, null);
```

查询数据库的所有数据。

在 Android 中查询数据是通过 Cursor 类来实现的，当使用 SQLiteDatabase.query()方法时，会得到一个 Cursor 对象，Cursor 指向的就是每一条数据。有很多有关查询的方法，请查阅 Android API 文档。

4.3.3 SQLiteOpenHelper 类

4.3.2 节介绍了 SQLite 数据库的相关操作方法。为了更加方便地管理、维护、升级数据库，Android API 提供了 SQLiteOpenHelper 类来管理 SQLite 数据库，通过继承这个类，实现一些方法就可以轻松地创建、打开以及操作数据库。

1．SQLiteOpenHelper

SQLiteOpenHelper 是一个辅助类，用来管理数据库的创建和版本。可以通过继承这个类，实现它的一些方法来对数据库进行一些操作。所有继承了这个类的类都必须实现下面这样的一个构造方法：

```
public SQliteopenHelper(Context context,String name,CursorFactory factory,int version)
```

context 是上下文对象，例如，一个 Activity。name 表示数据库文件名（不包括文件路径），SQLiteOpenHelper 类会根据这个文件名来创建数据库文件。factory 是一个可选的游标工厂（通常是 Null）。version 表示数据库的版本号。表 4-5 给出了 SQLiteOpenHelper 类的常用方法。

表 4-5 SQLiteOpenHelper 类常用方法

方　　法	描　　述
public void onCreate(SQLiteDatabase db);	第一次创建的时候调用
public void onUpgrade (SQLiteDatabase db, int oldVersion, int newVersion)	当数据库需要升级时调用
public void onOpen(SQLiteDatabase db)	打开数据库时调用
public SQLiteDatabase getReadableDatabase ()	以可读的方式创建或打开一个数据库
public SQLiteDatabase getWritableDatabase ()	以可写的方式创建或打开一个数据库

当要使用数据库的时候，SQLiteOpenHelper 类会检测数据库文件是否存在，如果数据库文件存在,则直接会打开数据库;如果不存在,则调用 SQLiteOpenHelper 类的 onCreate() 方法创建数据库。通过覆写 onCreate()、onUpgrade() 和 onOpen()对数据库进行操作。

2．SQLiteOpenHelper 的具体用法

创建一个新的 Class（见第 4 章 db.cqupt 包中的代码），如下所示，onCreate(SQLiteDatabase db)和 onUpgrade(SQLiteDatabase db，int oldVersion，int newVersion)方法会被自动添加。

代码具体如下：

```
1.  package db.cqupt;
2.  import android.content.Context;
3.  import android.database.sqlite.SQLiteDatabase;
4.  import android.database.sqlite.SQLiteOpenHelper;
5.
6.  public class DBconnection extends SQLiteOpenHelper {
7.      private final static int DATABASE_VERSION = 1;// 数据库版本号
8.      private final static String DATABASE_NAME="book.db";// 数据库名
9.      private static Context context;
10.
11.     public static void setContext(Context context) {
12.         DBconnection.context = context;
13.     }
14.
15.     public DBconnection() {
16.         super(context, DATABASE_NAME, null, DATABASE_VERSION);
17.     }
18.
19.     public void onCreate(SQLiteDatabase db) {
20.         String sql = "CREATE TABLE book(" + "id VARCHAR(30) NOT NULL, "      //"+" id中id是传入的参数
21.             + "name VARCHAR(30) NOT NULL, "
22.             + "price VARCHAR(30)NOT NULL) ;";
23.         db.execSQL(sql);                //执行SQL语句，在这里创建一个表
24.     }
25.
26.     public void onUpgrade(SQLiteDatabase db, int oldVersion, int newVersion) {
27.     public SQLiteDatabase getConnection() {
28.         SQLiteDatabase db=getWritableDatabase();
29.         return db;
30.     }
31.
32.     public void close(SQLiteDatabase db) {
```

```
33.         db.close();
34.     }
35.
36. }
```

第 19~24 行是建立一张名称为 book 的表，其中 "+" id 中的 id 是代表传入的参数，VARCHAR(30) 表示数据类型，NOT NULL 表示是否为空。在第 28 行通过 getWritableDatabase()方法获得 SQLiteDatabase 类的对象，然后对数据库进行操作。完成对数据库的操作后，在第 33 行关闭数据库。

4.3.4 数据库文件存储位置（SD 卡/手机内存）

前面讲解了 SQLite 数据库中的数据基本操作，现在来讲一下数据库文件存储的位置。数据库文件存储的位置有两种：一种是存储在默认系统/data/data/包名/databases 路径下，另一种则是保存在了/sdcard/文件夹 XXX 下，生成数据库文件 XXX.db。

数据库文件存储到 SD 卡中步骤如下：

第一步，确认模拟器存在 SD 卡，创建方法如下：

在 Eclipse 中创建虚拟器（AVD）时，可以指定 SD 卡的大小，如图 4-12 所示。

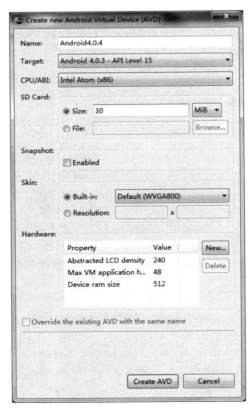

图 4-12 创建虚拟器时指定 SDCard 大小

第二步，先创建 SD 卡目录和路径。如果没有数据库，会默认以系统路径生成一个数据库和一个数据库文件，这里必须手动创建数据库文件。

第三步，在进行删除数据、添加数据等操作之前，通过打开第二步创建好的文件得到数据库实例。

第四步，创建表。

第五步，在配置文件 AndroidMainfest.xml 声明写入 SD 卡的权限。

4.4 数据共享 ContentProvider

前面说的 Preference、SQLite 等都是针对每个应用程序来说的，就是说数据不能跨越工程，只能供本工程使用，其他工程不能使用。但是有的时候程序是需要数据共享的（例如，Android 手机的短信和通讯录等都是数据共享的，对应的 ContentProvider 提供了这个接口，编写源码时就可以对其进行操作），这时就需要 ContentProviders 了，它是多个应用程序之间数据存储和检索的桥梁，主要作用就是在各个应用程序之间实现数据共享。

4.4.1 Android 系统自带的 ContentProvider

ContentProvider（数据提供者）是一种实现不同应用程序之间数据共享的接口机制，第 1 章已经讲过，ContentProvider 属于 Android 应用程序的组件之一，作为应用程序之间唯一的共享数据的途径，ContentProvider 主要的功能就是存储并检索数据以及向其他应用程序提供访问数据的接口。这里会详细讲解。我们知道，数据是私有的，只能被创建程序访问，其他应用程序是不能访问的。虽然前面已经讲过 Preference、文件系统的跨边界访问数据的用法，但是这些方法都有一定的局限性，它们有各自的缺点，而 ContentProvider 则提供了更加高级的方法实现数据共享，应用程序可以指定需要共享的数据，其他应用程序也可以在不知道数据来源、路径的情况下对数据进行增、删、改、查等操作。

共享数据的方式有两种：一是创建自己的 ContentProvier，二是将数据添加到已有的 ContentProvider 中。后者需要保证现有的 ContentProvider 和添加的数据类型相同且具有该 ContentProvider 的写入权限。对于 ContentProvider，实现数据共享最重要的就是数据模型（data model）和 URI。

1．数据模型

ContentProvider 将其存储的数据以数据表的形式提供给访问者，在数据表中每一行为一条记录，每一列为具有特定类型和意义的数据。每一条数据记录都包括一个"_ID"数值字段，该字段唯一标识一条数据。在几乎所有的 ContentProvider 的操作中都会用到 URI。

2．URI

每一个 ContentProvider 都对外提供一个能够唯一标识自身数据集（data set）的公开 URI，如果一个 ContentProvider 管理多个数据集，其将会为每个数据集分配一个独立的

URI。所有的 ContentProvider 的 URI 都以 "content://" 开头,其中 "content:" 用来标识数据是由 ContentProvider 管理的 schema。

在创建 ContentProvider 时需要先使用数据库、文件系统或者网络实现底层存储功能,然后再继承 ContentProvider 的类实现基本数据操作的接口函数(增加、删除、修改、查找)。但是调用者并不能直接使用 ContentProvider,而是使用 ContentResolver 对象与 ContentProvider 进行交互,ContentResolver 则通过 URI 确定需要访问的 ContentProvider,如图 4-13 所示。

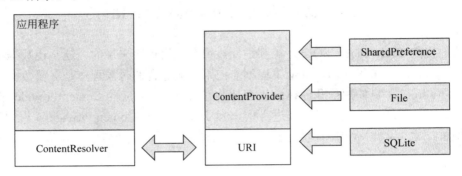

图 4-13　ContentProvider 流程

ContentProvider 完全屏蔽了数据提供组件的数据存储方法,使用者只知道 ContentProvider 提供了一种数据操作的接口机制,却不知道数据提供方是谁。数据提供方可以是 SQLite 或者是文件系统或者是 Preference,这些数据对使用者都是不可见的。这样就降低了 ContentProvider 的难度。只要调用这个接口函数就可以完成所有的数据操作了。

要对 ContentProvider 进行操作,一般需要通过 GetContentResolver()方法获得 ContentResolver 对象,然后知道唯一标识 ContentProvider 的 URI 以及调用 ContentResolver 中相应操作的方法。

4.4.2　自定义 ContentProvider

ContentProvider(内容提供器)用于与数据源打交道,而 ContentResolver(内容解析器)负责操作具体的 ContentProvider,数据源可以是文件或数据库。多个 ContentResolver 可以同时访问 ContentProvider,因为它的线程是安全的。一个 URI 可以确定一个资源,而一个 ContentProvider 可以有多个 URI,但所有 URI 的 Authority 相同。Android 没有共享内存,因此要访问另一个进程的数据,必须要通过 ContentProvider 来操作。

上面讲过与其他应用程序共享数据有两种方式:创建自己的 ContentProvier,即继承自 ContentProvider 的子类,或者是将自己的数据添加到已有的 ContentProvider 中,也就是说,ContentProvider 可以是系统的,也可以是自己创建的。现在我们来讲讲怎样创建 ContentProvider。

创建 ContentProvider 的步骤大致如下（具体的情况还要具体分析）：

（1）创建保存数据的文件或数据库。

（2）定义一个类继承 ContentProvider 实现抽象方法。

（3）在 Androidmainfest.xml 配置文件中声明定义好的 ContentProvider，以便于使用。

因为 ContentProvider 是与数据源打交道操作数据，所以必须要有保存数据的场所。可以使用 SQLite 数据库或者文件的方式保存数据，一般都是用 SQLite 数据库。所以要创建数据库来保存数据。要访问数据实现对数据的操作就要提供访问数据的接口，所以需要 ContentProvider 类，实现其中的抽象方法即增、删、改、查等。当然定义好的 ContentProvider 必须在 Androidmainfest.xml 中声明才可以使用。

4.5 一个有本地数据库的单机版图书管理系统

在 3.9 节，我们以一个有多个界面的单机版图书管理系统来对整个第 3 章所学知识进行实例说明。本章在第 3 章的有多个界面的单机版图书管理系统上进行扩充，为程序添加一个本地数据库 SQLite，这就是本章将要介绍的一个有本地数据库的单机版图书管理系统。

由于 Chapter04 是在 Chapter03 的基础上进行改动，所以整个思路与 Chapter03 相同，都是采用了 MVC 设计模式，两者不同的是：Chapter04 增加了一个 db.cqupt 包实现连接数据库。Chapter04 的包结构如图 4-14 所示。

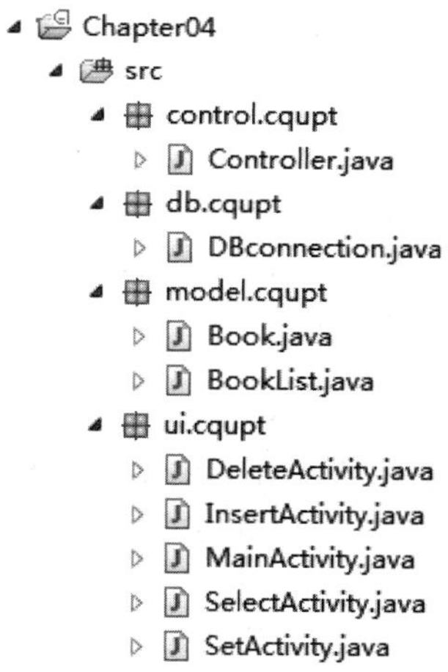

图 4-14　包结构

Chapter04 还在 res 下的 raw 文件夹中手动创建了一个数据库文件 book.db，如图 4-15 所示。

图 4-15　数据库文件

可以先在虚拟机上运行 Chapter04 看一看效果。下面将为读者详细讲述 Chapter04 的实现过程，Chapter04 的界面的布局等实现都与 Chapter03 相同，此处不再赘述。如果读者对界面还有疑问，请返回 3.10 节进行回顾。Chapter04 各个类之间的关系如图 4-16 所示。

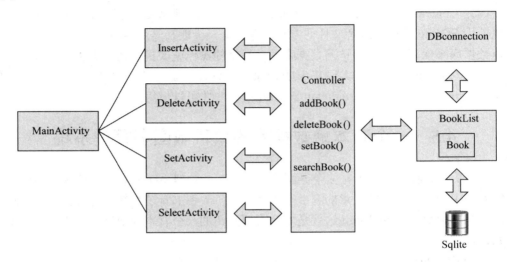

图 4-16　各个类之间的关系

可以与 Chapter03 中的图 3-44 对比，会发现只增加了 DBconnection 类。下面首先就对 DBconnection 类进行详细讲解，代码如下所示：

```
1.  package db.cqupt;
2.
3.  import android.content.Context;
4.  import android.database.sqlite.SQLiteDatabase;
5.  import android.database.sqlite.SQLiteOpenHelper;
6.
7.  public class DBconnection extends SQLiteOpenHelper {
8.      private final static int DATABASE_VERSION = 1;// 数据库版本号
9.      private final static String DATABASE_NAME = "book.db";// 数据库名
10.
11.     private static Context context;
12.
13.     public static void setContext(Context context) {
14.         DBconnection.context = context;
15.     }
16.
17.     public DBconnection() {
18.         super(context, DATABASE_NAME, null, DATABASE_VERSION);
19.     }
20.
```

```
21.     public void onCreate(SQLiteDatabase db) {
22.         String sql = "CREATE TABLE book(" + "id VARCHAR(30) NOT NULL,"
23.                 + "name VARCHAR(30) NOT NULL,"
24.                 + "price          VARCHAR(30) NOT NULL);";
25.         db.execSQL(sql);
26.     }
27.
28.     public void onUpgrade(SQLiteDatabase db,int oldVersion,int newVersion){
29.
30.     }
31.
32.     public SQLiteDatabase getConnection() {
33.         SQLiteDatabase db = getWritableDatabase();
34.         return db;
35.     }
36.
37.     public void close(SQLiteDatabase db) {
38.         db.close();
39.     }
40.
41. }
```

DBconnection 类的功能是对 SQLite 数据操作。首先 DBconnection 继承了 SQLiteOpenHelper 类。根据 4.3 节对 SQLite 数据库的讲解，我们已经知道，SQLiteOpenHelper 包装了 SQLite 数据库的创建、打开和更新的抽象类，通过实现和使用 SQLiteOpenHelper，可以隐去在数据库打开之前需要判断数据库是否需要创建或更新的逻辑。SQLiteOpenHelper 是一个抽象类，在该类中有两个抽象方法，SQLiteOpenHelper 的子类必须实现这两个方法：

```
public abstract void onCreate(SQLiteDatabase db);
public abstract void onUpdate(SQLiteDatabase db,int oldVersion,int newVersion);
```

首先在构造方法中分别需要传入 context 对象、数据库名称、CursorFactory（一般传入 null，为默认数据库）和数据库版本号。我们在第 11 行声明了一个静态的 context 对象，在第 13～15 行自定义了一个设置 context 对象的静态函数，这个函数会在主界面 MainActivity 中调用，MainActivity 中调用的代码如下所示（在 Chapter03 的 MainActivity 中的 onCreate()方法中插入的）：

```
DBconnection.setContext(this.getApplicationContext());
```

将整个应用程序的上下文变量（Context）作为参数传递给 setContext 方法来设置 DBconnection 中声明的 context 静态成员变量。代码第 8 行设置了数据库版本号，第 9 行设置了数据库名。在第 17～19 行的构造函数中传入各个参数。

代码第 21～26 行实现了 onCreate(SQLiteDatabase db)方法。在第 22～24 行书写了创建数据库表的 SQL 语句，第 27 行通过 SQLiteDatabase 对象的 execSQL()方法执行 SQL 语句。onCreate()方法会在数据库第一次生成的时候被调用。

代码第 28～30 行实现了 onUpdate(SQLiteDatabase db，int oldVersion，int newVersion)方法，当数据库需要升级的时候，Android 就会去调用 onUpdate()方法。这里不需要对数据库升级，所以这个方法为空。

代码第 32～35 行自定义了一个放回 SQLiteDatabase 对象的成员方法 getConnection，在这个方法中第 33 行调用了 getWritableDatabase()方法得到 SQLiteDatabase 对象并返回，可以通过这个 SQLiteDatabase 对象对 Sqlite 数据库进行操作。第 37～39 行是关闭数据库的方法，通过调用 SQLiteDatabase 对象的 close()方法关闭。

至此已对 DBconnection 的内容做了详细讲解，下面将讲述程序在何处使用了 DBconnection 和如何对 SQLite 数据库操作。如图 4-16 所示，可以知道是通过 BookList 类对 SQLite 数据库操作。下面详细讲述 BookList 中的代码，代码如下所示：

```
1.  package model.cqupt;
2.
3.  import java.util.ArrayList;
4.
5.  import db.cqupt.DBconnection;
6.  import android.database.Cursor;
7.  import android.database.sqlite.SQLiteDatabase;
8.
9.  public class BookList extends ArrayList<Book> {
10.
11.     private static final long serialVersionUID = 1L;
12.
13.     private static BookList booklist = null;
14.
15.     private BookList() {
16.
17.     }
18.
19.     public static BookList getBookList() {
20.         if (booklist == null) {
21.             booklist = new BookList();
22.             DBconnection connection = new DBconnection();
23.             SQLiteDatabase db = connection.getConnection();
24.             Cursor cur = db.query("book", null, null, null, null,
                    null, null);
25.             while (cur.moveToNext()) {
26.                 int idNum = cur.getColumnIndex("id");
27.                 int nameNum = cur.getColumnIndex("name");
```

```
28.                int priceNum = cur.getColumnIndex("price");
29.                String id = cur.getString(idNum);
30.                String name = cur.getString(nameNum);
31.                String price = cur.getString(priceNum);
32.                Book book = new Book(id, name, price);
33.                booklist.add(book);
34.                cur.moveToNext();
35.            }
36.            connection.close(db);
37.        }
38.        return booklist;
39.
40.    }
41.
42.    /**
43.     * 查找id
44.     */
45.    private boolean checkId(String bookid) {
46.        for (int i = 0; i < booklist.size(); ++i) {
47.            Book book = booklist.get(i);
48.            String id = book.getId();
49.            if (id.equals(bookid)) {
50.                return true;
51.            }
52.        }
53.        return false;
54.    }
55.
56.    /**
57.     * 得到图书位置
58.     */
59.    private int getIndex(String bookid) {
60.        int i = 0;
61.        for (; i < booklist.size(); ++i) {
62.            Book book = booklist.get(i);
63.            String id = book.getId();
64.            if (id.equals(bookid)) {
65.                break;
66.            }
67.        }
68.        return i;
69.    }
70.
71.    /**
```

```
72.        * 查找名称
73.        */
74.       public boolean checkName(String name) {
75.           for (int i = 0; i < booklist.size(); ++i) {
76.               Book book2 = booklist.get(i);
77.               if (book2.getName().equals(name)) {
78.                   booklist.remove(i);
79.                   return true;
80.               }
81.           }
82.           return false;
83.       }
84.
85.       public boolean insert(Book book) {
86.
87.           if (checkId(book.getId())) {
88.               return false;
89.           } else {
90.               booklist.add(book);
91.               String id = book.getId();
92.               String name = book.getName();
93.               String price = book.getPrice();
94.               DBconnection connection = new DBconnection();
95.               SQLiteDatabase db = connection.getConnection();
96.               String sql = "INSERT INTO book(id,name,price)" + "VALUES('" + id
97.                       + "','" + name + "','" + price + "')";
98.               db.execSQL(sql);
99.               db.close();
100.              return true;
101.          }
102.      }
103.
104.      public boolean delete(String name) {
105.          if (checkName(name)) {
106.              DBconnection connection = new DBconnection();
107.              SQLiteDatabase db = connection.getConnection();
108.              String sql = "DELETE FROM book WHERE name='" + name
                          + "'";
109.              db.execSQL(sql);
110.              return true;
111.          } else {
112.              return false;
113.          }
```

```
114.            }
115.
116.         public boolean set(Book book) {
117.             if (checkId(book.getId())) {
118.                 String id = book.getId();
119.                 String name = book.getName();
120.                 String price = book.getPrice();
121.                 int index = getIndex(id);
122.                 booklist.set(index, book);
123.                 DBconnection connection = new DBconnection();
124.                 SQLiteDatabase db = connection.getConnection();
125.                 String sql = "UPDATE book SET name='" + name + "',"
                         + "price='"
126.                         + price + "' WHERE id='" + id + "'";
127.                 db.execSQL(sql);
128.                 return true;
129.             } else {
130.                 return false;
131.             }
132.         }
133.
134.
135.    }
```

在 Chapter03 中 BookList 类的功能是存储图书，而在 Chapter04 中，新增加了对 SQLite 数据库实现增、删、改、查的功能，使 booklist 存储的图书和 SQLite 数据库中的图书同步。

增加图书功能在第 85～102 行的 insert 方法中实现。首先在第 87 行通过 checkId 判断即将增加的图书编号在 booklist 对象中是否存在（booklist 与 SQLite 同步），在第 45～54 行通过循环遍历 booklist 对象，判断是否有相同编号的图书，如果有，则返回 true；否则返回 false。insert 函数通过 checkId 的返回值判断后续执行的代码，如果 checkId 返回 true，那么 insert 返回 false 代表插入失败；否则执行第 90～100 行对 SQLite 数据库操作，第 90 行在 booklist 对象中添加这本图书，第 91～93 行得到图书的编号、名称、价格。第 94 行得到 connection 对象并且调用 connection 对象的 getConnection 方法得到 SQLiteDatabase 对象 db，通过 db 对 SQLite 数据库进行增加操作，第 96、97 行是插入图书的 SQL 语句，第 109～110 行执行 SQL 语句，关闭数据库并且返回 true。

删除图书的功能是在第 104～114 行的 delete 方法中实现的。通过传递的图书名称删除数据库的图书。首先在第 105 行通过 checkName 方法判断即将删除的图书名称在 booklist 中是否存在。在第 74～83 行的 checkName 方法中，通过循环变量 booklist 中的图书名称判断是否有相同名称的图书，如果有，则通过第 78 行 booklist 的 remove 方法删除 booklist 中的这本图书，并且返回 true；否则返回 false。delete 方法根据 checkName

的返回值执行相应的代码，如果 checkName 返回 true，则执行第 106～110 行，和 insert 方法一样，通过执行删除图书的 SQL 语句操作 SQLite 数据库并且返回 true。

修改图书的功能是在第 116～132 行的 set 方法中实现的。首先在第 117 行通过 checkId 判断即将增加的图书编号在 booklist 对象中是否存在，然后根据 checkId 的返回值判断后续执行的代码，如果 checkId 返回 true，则执行第 118～128 行，得到修改图书信息并且更新数据库和 booklist 并且返回 true，在第 121、122 行是对 booklist 的操作，通过 getIndex 方法得到要修改图书在 booklist 中的位置，然后使用 booklist 的 set 方法更新这个位置的图书。

查询图书的功能是通过得到 booklist 对象（得到 booklist 对象的方法在第 19～40 行的 getBookList 静态方法中）实现的。SQLite 数据库中的数据通过 getBookList 方法存入到 booklist 对象中，这部分代码运用了前面学习的查询数据库的知识，第 20 行首先判断 booklist 对象是否被创建了，如果已经创建了则直接返回 booklist 对象，否则执行第 21 行创建 booklist 对象，在第 22、23 行得到操作数据库的对象，在第 24～35 行是查询 SQLite 数据库的方法，通过 Cursor 对象遍历 SQLite 数据库，将得到的数据插入到 booklist 中，最后在第 38 行返回 booklist 对象。

由图 4-16 可知，BookList 被 Controller 操控，下面是 Controller 类的代码：

```
1.  package control.cqupt;
2.
3.  import model.cqupt.Book;
4.  import model.cqupt.BookList;
5.
6.  public class Controller {
7.      public boolean addBook(String id, String name, String price) {
8.          BookList booklist = BookList.getBookList();
9.          Book book = new Book(id, name, price);
10.         if (booklist.insert(book))
11.             return true;
12.         else
13.             return false;
14.     }
15.
16.     public boolean deleteBook(String name) {
17.         BookList booklist = BookList.getBookList();
18.         if (booklist.delete(name))
19.             return true;
20.         else
21.             return false;
22.     }
23.
24.     public boolean setBook(String id, String name, String price) {
25.         BookList booklist = BookList.getBookList();
```

```
26.            Book book = new Book(id, name, price);
27.            if (booklist.set(book))
28.                return true;
29.            else
30.                return false;
31.        }
32.
33.        public BookList searchBook() {
34.            BookList booklist = BookList.getBookList();
35.            return booklist;
36.
37.        }
38.
39. }
```

Chapter04 的 Controller 类中的代码和 Chapter03 的 Controller 代码基本上相同，从这点读者也可以感受到 MVC 设计模式的好处。在这里就不对 Controller 类进行讲解了。由于 Chapter04 的 ui.cqupt 包中的所有界面 Activity 类和 Chapter03 的实现基本相同，所以读者可以自行查阅本章节的相关代码。

第 5 章 网 络 编 程

Java 语言提供了丰富的网络编程类库。查看 Android API 文档会发现，Android 支持 Java 提供的所有网络编程方式，所以本章通过具体实例对 Java 网络编程进行讲解，循序渐进地让读者理解网络编程。5.2 节将从最简单的控制台程序开始学起，掌握最简单的 IO 操作。从实际教学情况看，可以发现几乎每个学过 Java 的同学都知道 System.out.println(…)，但并不是每个同学都知道到怎么实现由键盘读入。可以发现，从键盘、显示器的默认输入输出开始，一旦建立起网络连接，接下来网络两端的读写操作和本地读写是一样的，本章的教学顺序、教学方法在实际教学中取得了较好的效果。

5.1 什么是网络编程

什么是网络编程？这是我们遇到的第一个问题。也是我在本科学习阶段一直很困惑的问题。

首先简单地回答网络为什么要分层这个问题。当一个问题比较复杂的时候，常见有两种方式处理。一是横向分。例如下课后我们要办个班级小聚会。张三同学带着几个同学准备布置场地、李四同学和几个同学去买吃的等。这样的协作形式就是横向的。落实到软件领域，例如做一个简单的网站，各个模块之间没有耦合或耦合很少，大家横向分工，各自之间互不干扰。二是纵向分，例如在淘宝上网购，从下订单、商家发货、物流、买家收货、淘宝确认等，这一系列的过程是承前启后、彼此连贯的，这就是分层。这里就不再扩展了。当然，在软件工程领域还有其他的一些方法，如原型法等，这里也不扩展了。

一般计算机网络课程中是按照 5 层的形式教学的，即物理层、数据链路层、网络层、传输层、应用层。每一层都会提供一些函数供我们调用。这在现实世界是一样的，例如去物流快递公司寄包裹，那么我们要填一个单子，这就是接口，或者说是函数调用。

我们学习的是 Android 应用程序开发，因此站在应用层。试想我们站在应用层，那么看到的有两层：传输层和应用层本身。因此，本章重点讲两种编程：一是调用传输层提供的 Socket 接口进行网络编程；二是调用应用层的函数、类接口进行网络编程。在 Socket 编程中，一般有两种：TCP 和 UDP，本章重点介绍 TCP。在应用层网络编程中有很多协议，重点学习最常使用的 HTTP。

5.1.1 Socket 通信

在 TCP/IP 网络应用中，通信的两个进程间相互作用的主要模式是客户/服务器模式（Client/Server model），即客户向服务器发出服务请求，服务器接收到请求后，提供相应的服务。TCP 通信中与 Socket 通信有两个相关的类：一个是代表客户端套接字的 Socket 类，另一个是代表服务器端的套接字的 ServerSocket 类。

1. 客户端

Socket 是对 TCP/IP 协议的封装，它本身并不是协议，而是一个调用接口（API），客户端通过构造一个 Socket 类对象建立与服务器的 TCP 连接。Socket 类的构造方法有如表 5-1 所示的 6 种。

表 5-1 Socket 类的构造方法

构 造 方 法	功 能 说 明
Socket()	创建套接字，不请求任何连接
Socket(InetAddress address, int port)	创建一个流套接字并将其连接到指定 IP 地址的指定端口号
Socket(InetAddress address, int port, InetAddress localAddr, int localPort)	创建一个套接字并将其连接到指定远程端口上的指定远程地址
Socket(Proxy proxy)	不管其他设置如何，都应使用的指定代理类型（如果有），根据创建一个未连接的套接字
Socket(SocketImpl impl)	创建带有用户指定的 SocketImpl 的未连接 Socket
Socket(String host, int port)	创建一个流套接字并将其连接到指定主机上的指定端口号

5.2 节讲解的实例中用到的是 Socket(String host, int port)这个构造方法建立连接，如 Socket("localhost","8008")，这意味着向本机发送请求通过 8008 端口连接，如果客户端发出的请求被服务器拒绝，Socket 构造方法就会抛出 ConnectionException。成功创建 Socket 对象后，通过 Socket 类提供的一系列方法，与服务器进行交互，Socket 类的主要方法如表 5-2 所示。

表 5-2 Socket 类的主要方法

方 法 名	功 能 说 明
void close()	关闭 Socket 连接
InputStream getInputStream()	获取 Socket 输入流
OutputStream getOutputStream()	获取 Socket 输出流
int getPort()	获取远程主机端口号
InetAddress getLocalAddress()	获取本地主机的 Internet 地址

客户端的实例将在 5.2 节详细讲述。

2. 服务器端

ServerSocket 类用来在服务器端监听所有来自指定接口的连接，并为每个新的连接创建一个 Socket 对象，客户端和服务器端就可以通过 Socket 对象进行通信了。ServerSocket 的构造方法如表 5-3 所示。

表 5-3　Server Socket 的构造方法

构 造 方 法	功 能 说 明
ServerSocket()	创建非绑定服务器套接字
ServerSocket(int port)	创建绑定到特定端口的服务器套接字
ServerSocket(int port, int backlog)	利用指定的 backlog 创建服务器套接字并将其绑定到指定的本地端口号
ServerSocket(int port, int backlog, InetAddress bindAddr)	使用指定的端口、侦听 backlog 和要绑定到的本地 IP 地址创建服务器

在 5.2 节的服务器端实例中，采用了 ServerSocket(8008)这个构造方法开启 8008 端口监听客户端连接，并且通过 ServerSocket 类的 accept()方法侦听并接收来自客户端的连接请求，返回 Socket 对象；如果没有客户端连入，则服务器端一直在 accept()方法这里处于阻塞状态。

5.1.2　HTTP 通信

在 Java 中使用 HTTP 通信的客户端需要用到 java.net 包中的 HttpURLConnection 类，每个 HttpURLConnection 实例都可用于生成单个请求，请求完成后在 HttpURLConnection 的 InputStream 或 OutputStream 上调用 close() 方法以释放与此实例关联的网络资源。表 5-4 是 HttpURLConnection 的常用方法。

表 5-4　HttpURLConnection 的常用方法

方 法	功 能 说 明
HttpURLConnection(URL u)	构造方法
String getRequestMethod(String method)	获取请求方法
int getResponseCode()	从 HTTP 响应消息获取状态码

在 HttpURLConnection 构造方法中通过 URL 对象指明要访问的 URL 地址。URL 在 java.net 包中，表 5-5 是 URL 的常用方法。

表 5-5　URL 的常用方法

方 法	功 能 说 明
URL(String spec)	根据 String 表示形式创建 URL 对象
String getHost()	获得此 URL 的主机名
URLConnection openConnection()	返回一个 URLConnection 对象，它表示到 URL 所引用的远程对象的连接
InputStream openStream()	打开到此 URL 的连接并返回一个用于从该连接读入的 InputStream

服务器端通过 Servlet 接收并响应客户端的请求，详细示例见 5.2.7 节。

5.2　客户/服务器模式

本节采用 7 个实例介绍从最简单的控制台输入输出到网络上通过 Socket 建立 C/S 模式的通信。C/S 模式采用由简单到复杂的思路：一个客户端和一个服务器端进行一次通信、一个客户端和一个服务器端进行多次通信、多个客户端和一个服务器端进行通信。最后讲解客户端和服务器端通过 HTTP 协议进行通信。

5.2.1 控制台上的简单输入输出

本节实现了一个在控制台上输入并且输出结果的示例,代码在工程 Chap5.2.1 中,运行本工程,在控制台中输入 hello 字符,如图 5-1 所示。

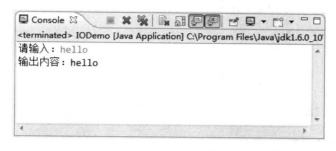

图 5-1　Chap5.2.1 执行结果

Chap5.2.1 代码如下所示:

```
1.  package io;
2.  import java.util.Scanner;
3.
4.  public class IODemo {
5.
6.      public IODemo() {
7.          System.out.print("请输入:");
8.          Scanner scanner = new Scanner(System.in);
9.          String str = scanner.next();
10.         System.out.println("输出内容:"+str);
11.     }
12.
13.     public static void main(String[] args) {
14.         new IODemo();
15.     }
16.
17. }
```

在第 8 行通过 java.util 包中的 Scanner 类实现控制台的输入输出,在第 9 行通过 Scanner 类的 next()方法得到输入到控制台中的字符串,并在第 10 行通过 System.out 输出字符串到控制台。

5.2.2 控制台上的循环输入输出

本节的工程实现在控制台上的循环输入输出,在 Chap5.2.1 工程的基础上加入了一个 while 循环，使程序可以循环输入输出。当输入 exit 字符串时，程序退出循环。运行本工程，在控制台中输入 hello 字符，效果如图 5-2 所示。

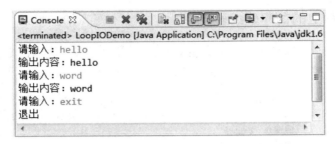

图 5-2　Chap5.2.2 执行结果

工程 Chap5.2.2 代码如下所示：

```
1.  package io;
2.  import java.util.Scanner;
3.
4.  public class LoopIODemo {
5.
6.      public LoopIODemo() {
7.          while (true) {
8.              System.out.print("请输入：");
9.              Scanner scanner = new Scanner(System.in);
10.             String str = scanner.next();
11.             if (str.equals("exit")) {
12.                 System.out.println("退出");
13.                 break;
14.             } else {
15.                 System.out.println("输出内容：" + str);
16.             }
17.         }
18.     }
19.
20.     public static void main(String[] args) {
21.
22.         new LoopIODemo();
23.     }
24.
25. }
```

本程序与 Chap5.2.1 相比，在第 7 行添加了一个 while 循环实现在控制台循环输入的功能，每次在控制台上输入的数据都会在第 11～13 行的条件语句中进行判断，如果输入的字符是 exit，则执行第 12、13 行，退出 while 循环，终止程序。

5.2.3　一个客户端和一个服务器端一次通信

从本节起开始介绍客户端与服务器端的 Socket 通信。首先是最简单的一个客户端和一个服务器端进行一次通信。工程 Chap5.2.3 中有两个包：client 包和 server 包。在 client 包中是客户端 SocketClient，在 server 包中是服务器端 SocketServer。首先运行服务器端的 SocketServer 类，没有连接客户端时的效果如图 5-3 所示。

图 5-3　服务器未连客户端时

此时没有客户端连入，所以服务器端一直等待客户端的连入。当运行 SocketClient 类时，服务器端的控制台运行效果如图 5-4 所示，客户端的控制台运行效果如图 5-5 所示。

图 5-4　服务器端

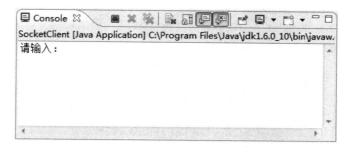

图 5-5　客户端

在客户端的控制台中输入如图 5-6 所示的数据并发送给服务器端，服务器端接收数据并返回一个字符串，然后断开与客户端的连接，如图 5-7 所示。

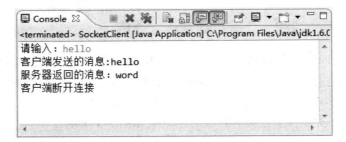

图 5-6　客户端发送消息

图 5-7　服务器端接收消息并返回

下面讲解 Chap5.2.3 的代码。首先是客户端 SocketClient，代码如下所示：

```
1.  package client;
2.  import java.io.IOException;
3.  import java.io.InputStream;
4.  import java.io.OutputStream;
5.  import java.net.Socket;
6.  import java.net.UnknownHostException;
7.  import java.util.Scanner;
8.
9.  public class SocketClient {
10.     public SocketClient() {
11.         Socket socket = null;
12.         OutputStream out = null;
13.         InputStream in = null;
14.         try {
15.             socket = new Socket("localhost", 8008);
16.             out = socket.getOutputStream();
17.             System.out.print("请输入: ");
18.             Scanner scanner = new Scanner(System.in);
19.             String mes = scanner.next();
20.             System.out.println("客户端发送的消息:" + mes);
21.             out.write(mes.getBytes());
22.             in = socket.getInputStream();
```

```
23.            byte[] buffer = new byte[1024];
24.            int index = in.read(buffer);
25.            String receive = new String(buffer, 0, index);
26.            System.out.println("服务器返回的消息: " + receive);
27.            System.out.println("客户端断开连接");
28.            in.close();
29.            out.close();
30.            socket.close();
31.        } catch (UnknownHostException e) {
32.            e.printStackTrace();
33.        } catch (IOException e) {
34.            e.printStackTrace();
35.        }
36.    }
37.
38.    public static void main(String[] args) {
39.        new SocketClient();
40.
41.    }
42.
43. }
```

客户端 SocketClient 在包 client 中，首先在第 15 行通过 Socket 连接本地监听端口号为 8008 的服务器端，在第 16 行和第 22 行得到 Socket 的输出输入流，通过得到的输出对象 out 向服务器端发送控制台中输入的消息，然后再通过输入对象 in 接收来自服务器端的返回消息，最后在第 28～30 行关闭各个连接。

下面来看一看服务器端 SocketServer 是如何实现的，代码如下所示：

```
1.  package server;
2.  import java.io.IOException;
3.  import java.io.InputStream;
4.  import java.io.OutputStream;
5.  import java.net.ServerSocket;
6.  import java.net.Socket;
7.
8.  public class SocketServer {
9.
10.     public SocketServer() {
11.         Socket socket = null;
12.         OutputStream out = null;
13.         InputStream in = null;
14.         try {
15.             ServerSocket serverscoket = new ServerSocket(8008);
16.             System.out.println("服务器等待客户端连接...");
```

```
17.             socket = serverscoket.accept();
18.             String ip = socket.getLocalAddress().getHostAddress();
19.             int port = socket.getPort();
20.             System.out.println("连接上的客户端ip: "+ip+",端口号: "+port);
21.             in = socket.getInputStream();
22.             byte[] buffer = new byte[1024];
23.             int index = in.read(buffer);
24.             String receive = new String(buffer, 0, index);
25.             System.out.println("服务器端接收到的消息: " + receive);
26.             out = socket.getOutputStream();
27.             String mes = "word";
28.             out.write(mes.getBytes());
29.             System.out.println("服务器发送的消息: " + mes);
30.             System.out.println("服务器断开连接");
31.             in.close();
32.             out.close();
33.             socket.close();
34.             serverscoket.close();
35.
36.         } catch (IOException e) {
37.             e.printStackTrace();
38.         }
39.
40.     }
41.
42.     public static void main(String[] args) {
43.         new SocketServer();
44.     }
45.
46. }
```

服务器端 SocketServer 在包 server 中，首先在第 15 行通过 ServerSocket 监听 8008 端口，通过 accept()方法接收客户端的连接，如果没有客户端接入，则程序一直阻塞在第 17 行，如图 5-3 所示。当客户端接入后，开始执行下面的代码通过 Socket 获取客户端的 IP 地址和端口号，并且得到与客户端进行通信的 Socket 输入输出流，完成和客户端之间的通信。服务器端在第 24 行得到了客户端发送的消息，立即在第 28 行返回一个字符串给客户端，然后关闭各个流，完成一次客户端和服务器端的通信。

本节讲述了客户端和服务器端的一次对话，在后面将介绍客户端与服务器端的进一步交互。

5.2.4 一个客户端和一个服务器端多次通信

5.2.3 节介绍了一个客户端和一个服务器端的一次通信，本节将对 5.2.3 节的功能进行扩充,让客户端能与服务器多次通信,直到发送退出消息结束通信。Chap5.2.4 在 Chap5.2.3 上进行了改动,修改了客户端 SocketClient,在 SocketClient 中增加了一个 while 循环,使客户端可以一直发送消息给服务器端,直到客户端发送退出消息。在服务器端的 SocketServer 中也增加了一个 while 循环,用于一直读取客户端发送的消息。下面来看一看 Chap5.2.4 的运行效果,首先分别运行服务器端和客户端,服务器端如图 5-8 所示。

图 5-8　服务器端

然后在客户端的控制台输入 hello 消息并发送给服务器端,客户端的控制台如图 5-9 所示。

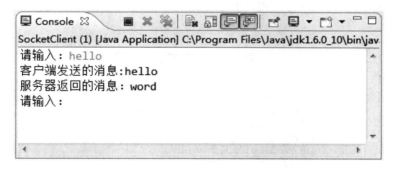

图 5-9　客户端发送一条消息

由图 5-9 可以看出,客户端还可以继续输入消息,那么再输入一条消息 hi,此时客户端的控制台如图 5-10 所示,服务器端如图 5-11 所示。

图 5-10　客户端发送第二条消息

图 5-11 服务器端接收了两条消息

最好当客户端发送 exit 消息时告诉服务器断开连接,客户端如图 5-12 所示,服务器端如图 5-13 所示。

图 5-12 客户端发送退出消息

图 5-13 服务器端接收到退出消息并退出

从以上各图中可以发现客户端与服务器端进行了多次通信,下面就来讲解如何实现此功能。首先讲解客户端 SocketClient,代码如下所示:

```
1.  package client;
2.
3.  import java.io.IOException;
4.  import java.io.InputStream;
5.  import java.io.OutputStream;
6.  import java.net.Socket;
```

```java
7.  import java.net.UnknownHostException;
8.  import java.util.Scanner;
9.
10. public class SocketClient {
11.     public SocketClient() {
12.         InputStream in =null;
13.         OutputStream out =null;
14.         Socket socket = null;
15.         try {
16.             socket=new Socket("localhost",8008) ;
17.             out = socket.getOutputStream();
18.             while(true)
19.             {
20.                 System.out.print("请输入: ");
21.                 Scanner scanner = new Scanner(System.in);
22.                 String mes = scanner.next();
23.                 System.out.println("客户端发送的消息:"+mes);
24.                 out.write(mes.getBytes());
25.                 if(mes.equals("exit"))break;
26.                 in = socket.getInputStream();
27.                 byte[] buffer = new byte[1024];
28.                 int index = in.read(buffer);
29.                 String receive = new String(buffer, 0, index);
30.                 System.out.println("服务器返回的消息: "+receive);
31.             }
32.             System.out.println("客户端断开连接");
33.             in.close();
34.             out.close();
35.             socket.close();
36.         } catch (UnknownHostException e) {
37.             e.printStackTrace();
38.         } catch (IOException e) {
39.             e.printStackTrace();
40.         }
41.     }
42.
43.     public static void main(String[] args) {
44.         new SocketClient();
45.
46.     }
47.
48. }
```

客户端 SocketClient 在包 client 中，将本代码和 Chap5.2.3 工程中的 SocketClient 对

比，可以发现，我们将第 20～30 行（客户端与服务器进行消息发送和接收）这段代码放入了一个 while 循环中，这样客户端就可以一直发送消息和获取消息，然后在第 25 行对客户端发送的消息进行判断，如果客户端发送了 exit（退出）消息，则客户端跳出循环，结束程序。

下面来看一看服务器端 SocketServer，代码如下所示：

```
1.  package server;
2.  import java.io.IOException;
3.  import java.io.InputStream;
4.  import java.io.OutputStream;
5.  import java.net.ServerSocket;
6.  import java.net.Socket;
7.
8.  public class SocketServer {
9.
10.     public SocketServer() {
11.         InputStream in =null;
12.         OutputStream out =null;
13.         Socket socket = null;
14.         try {
15.             ServerSocket serverscoket = new ServerSocket(8008);
16.             System.out.println("服务器等待客户端连接...");
17.             socket = serverscoket.accept();
18.             String ip = socket.getLocalAddress().getHostAddress();
19.             int port = socket.getPort();
20.             System.out.println("连接上的客户端ip: " + ip + ",端口号: " + port);
21.             while(true)
22.             {
23.                 in = socket.getInputStream();
24.                 byte[] buffer = new byte[1024];
25.                 int index = in.read(buffer);
26.                 String receive = new String(buffer, 0, index);
27.                 System.out.println("服务器端接收到的消息: " + receive);
28.                 if(receive.equals("exit"))break;
29.                 out = socket.getOutputStream();
30.                 String mes = "word";
31.                 out.write(mes.getBytes());
32.                 System.out.println("服务器发送的消息: "+mes);
33.             }
34.             System.out.println("服务器断开连接");
35.             in.close();
36.             out.close();
```

```
37.                socket.close();
38.                serverscoket.close();
39.
40.            } catch (IOException e) {
41.                e.printStackTrace();
42.            }
43.
44.     }
45.
46.     public static void main(String[] args) {
47.         new SocketServer();
48.     }
49.
50. }
```

服务器端 SocketServer 在 server 包中，同样，我们把本代码和 Chap5.2.3 的 SocketServer 对比，可以发现，本代码只是在 Chap5.2.3 的 SocketServer 收发消息这一部分增加了一个 while 循环，使服务器端可以不停地接收和响应客户端信息。在第 28 行对收到的消息进行判断，如果收到的是客户端的退出消息，则服务器端也退出循环并且关闭连接。

本节实现了一个客户端和一个服务器端的多次通信，那么多个客户端和一个服务器端应该如何实现呢？

5.2.5 多个客户端和一个服务器端串行通信

5.2.4 节讲述了一个客户端和一个服务器端的多次通信，本节将在 5.2.4 节的 Chap5.2.4 的基础上对服务器端代码进行修改，再加入一个 while 循环，使服务器端接收多个客户端的消息。

首先运行服务端，并且运行一个客户端 client_1，服务器端连入一个客户端，如图 5-14 所示。

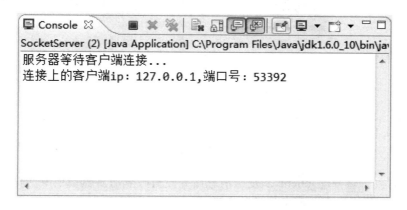

图 5-14 服务器端连入一个客户端

然后客户端 client_1 发送一条消息给服务器端，服务器端如图 5-15 所示。

图 5-15　服务器端

此时再次运行一个客户端 client_2，此时服务器端如图 5-16 所示。

图 5-16　启动 client_2 后

可以发现客户端 client_2 没有和服务器端建立连接。为了进一步验证，用客户端 client_2 服务器端发送一条消息，客户端 client_2 如图 5-17 所示。

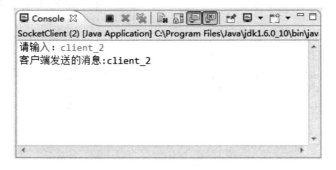

图 5-17　客户端 client_2

可以看到，客户端 client_2 发送消息后并没有收到服务器端的响应信息，证明了客户端 client_2 确实没有和服务器端建立连接。下面将断开客户端 client_1 与服务器端的连接，如图 5-18 所示。

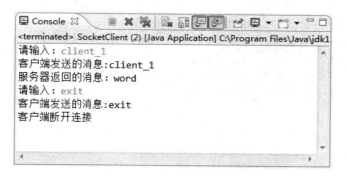

图 5-18　client_1 断开连接

此时服务器端和客户端 client_2 发生了变化，服务器端如图 5-19 所示，客户端 client_2 如图 5-20 所示。

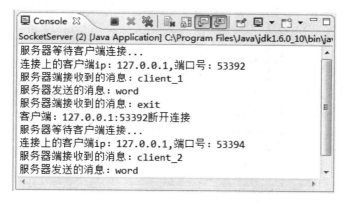

图 5-19　服务器端

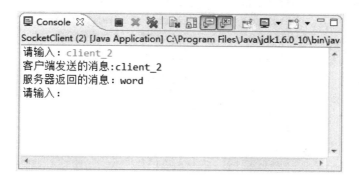

图 5-20　客户端 client_2

可以看到，当客户端 client_1 断开与服务器端的连接后，服务器端立即与客户端 client_2 建立连接，收到客户端 client_2 的消息并返回响应消息。

服务器端 SocketServer 代码如下所示：

```
1.  package server;
```

```
2.   import java.io.IOException;
3.   import java.io.InputStream;
4.   import java.io.OutputStream;
5.   import java.net.ServerSocket;
6.   import java.net.Socket;
7.
8.   public class SocketServer {
9.
10.      public SocketServer() {
11.          InputStream in =null;
12.          OutputStream out =null;
13.          Socket socket = null;
14.          try {
15.              ServerSocket serverscoket = new ServerSocket(8008);
16.              while(true)
17.              {
18.                  System.out.println("服务器等待客户端连接...");
19.                  socket = serverscoket.accept();
20.                  String ip = socket.getLocalAddress().getHostAddress();
21.                  int port = socket.getPort();
22.                  System.out.println("连接上的客户端ip: " + ip + ",端口号: " + port);
23.                  while(true)
24.                  {
25.                  in = socket.getInputStream();
26.                  byte[] buffer = new byte[1024];
27.                  int index = in.read(buffer);
28.                  String receive = new String(buffer, 0, index);
29.                  System.out.println("服务器端接收到的消息: " + receive);
30.                  if(receive.equals("exit"))break;
31.                  out = socket.getOutputStream();
32.                  String mes = "word";
33.                  out.write(mes.getBytes());
34.                  System.out.println("服务器发送的消息: "+mes);
35.                  }
36.                  System.out.println("客户端: "+ip+":"+port+"断开连接");
37.                  in.close();
38.                  out.close();
39.              }
40.
41.          } catch (IOException e) {
42.              e.printStackTrace();
43.          }
44.
```

```
45.     }
46.
47.     public static void main(String[] args) {
48.         new SocketServer();
49.     }
50.
51. }
```

服务器端 SocketServer 在 server 包中，在 Chap5.2.4 的基础上在第 16 行新增加了一个 while 循环，将服务器端等待客户端的连接这段代码放入 while 循环中。通过以上实例可以看出，服务器端是没有办法在与客户端 client_1 通信的时候同时等待其他客户端的连接和通信的，只有当客户端 client_1 退出后，服务器端才能与客户端 client_2 进行通信。

那么怎么才能实现服务器端一边等待其他客户端的通信，一边和已连上的客户端进行通信呢？5.2.6 节将对此做介绍。

5.2.6 多个客户端和一个服务器端并行通信

5.2.5 节想要实现多个客户端和一个服务器端多次通信，但遇到了一个问题：服务器端不能在等待其他客户端的连入的同时和已连入的客户端通信，本节通过 Java 多线程解决这个问题。

首先运行 Chap5.2.6 的服务器端 SocketServer 和客户端 client_1，如图 5-21 所示。

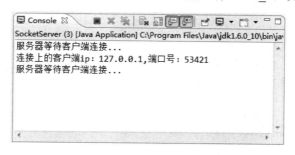

图 5-21　接入一个客户端

再次打开一个客户端 client_2，此时观察服务器端的控制台，如图 5-22 所示。

图 5-22　接入两个客户端

和 5.2.5 节的 Chap5.2.5 运行结果相比较，可以发现本工程的服务器端可以接受两个客户端同时连接，也可以与客户端 client_1 和客户端 client_2 进行通信，而不需要关闭其中一个客户端，如图 5-23 所示。

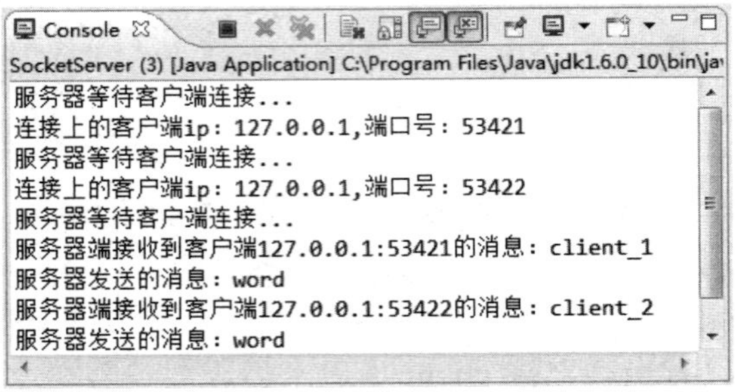

图 5-23　与不同客户端通信

下面来看一下如何实现本工程。本工程在 Chap5.2.4 的基础上对服务器端进行了修改，当每次有客户端连入时，都会在服务器端为这个客户端开一个独立的线程收发消息。服务器端 SocketServer 代码如下所示：

```
1.  package server;
2.  import java.io.IOException;
3.  import java.net.ServerSocket;
4.  import java.net.Socket;
5.
6.  public class SocketServer {
7.
8.      public SocketServer() {
9.          Socket socket = null;
10.         try {
11.             ServerSocket serverscoket = new ServerSocket(8008);
12.             while(true)
13.             {
14.                 System.out.println("服务器等待客户端连接...");
15.                 socket = serverscoket.accept();
16.                 String ip = socket.getLocalAddress().getHostAddress();
17.                 int port = socket.getPort();
18.                 System.out.println("连接上的客户端ip: " + ip + ",端口号: " + port);
19.                 new ServerThread(socket).start();
20.             }
21.
22.         } catch (IOException e) {
23.             e.printStackTrace();
24.         }
```

```
25.
26.     }
27.
28.     public static void main(String[] args) {
29.         new SocketServer();
30.     }
31. }
```

与 Chape5.2.4 的 SocketServer 对比,在第 19 行新增加了一个开启线程的功能,服务器端通过第 12 行的 while 循环不断接收客户端的连入请求,当客户端连入后,在第 19 行为连入的客户端开启一个新的线程用于与客户端进行通信,此时服务器端又回到第 15 行继续等待下一个客户端的连入请求。我们就是通过开启新线程的方式解决了工程 Chap5.2.5 服务器端不能同时接收多个客户端的连入请求问题的。线程 ServerThread 代码如下所示:

```
1.  package server;
2.  import java.io.IOException;
3.  import java.io.InputStream;
4.  import java.io.OutputStream;
5.  import java.net.Socket;
6.
7.  public class ServerThread extends Thread {
8.      private Socket socket;
9.
10.     public ServerThread(Socket socket) {
11.         this.socket = socket;
12.     }
13.
14.     public void run() {
15.         InputStream in = null;
16.         OutputStream out = null;
17.         String ip = socket.getLocalAddress().getHostAddress();
18.         int port = socket.getPort();
19.         try {
20.             while (true) {
21.                 in = socket.getInputStream();
22.                 byte[] buffer = new byte[1024];
23.                 int index = in.read(buffer);
24.                 String receive = new String(buffer, 0, index);
25.                 System.out.println("服务器端接收到客户端"+ip+":
                    "+port+"的消息: " + receive);
26.                 if (receive.equals("exit"))
27.                     break;
28.                 out = socket.getOutputStream();
```

```
29.              String mes = "word";
30.              out.write(mes.getBytes());
31.              System.out.println("服务器发送的消息: " + mes);
32.          }
33.          System.out.println("客户端: "+ip+":"+port+"断开连接");
34.          in.close();
35.          out.close();
36.          socket.close();
37.      } catch (IOException e) {
38.          e.printStackTrace();
39.      }
40.
41.     }
42. }
```

ServerThread 在 server 包中，ServerThread 是一个线程类，因为它继承了 Thread（注意：实现了 Runnable 接口的类也是线程类），并且覆写了 run 方法，在 run 方法中实现了与客户端之间的通信，实现通信的方法和 Chap5.2.4 是一样的，此处不再赘述。

5.2.7　客户端与服务器端 HTTP 通信

前面几节介绍了客户端与服务器端的 TCP 通信方式，通过 Socket 连接，本节讲述 HTTP 通信，Chap5.2.7 使用 tomcat 作为服务器。下面运行 Chap5.2.7 的服务器端 HttpServer 和客户端 HttpClient，并且在客户端输入 hello 字符串，客户端如图 5-24 所示，服务器端如图 5-25 所示。

图 5-24　HTTP 客户端

图 5-25　HTTP 服务器端

HTTP 和 TCP 的区别是：HTTP 在每次请求结束后都会主动释放连接，因此 HTTP 连接是一种"短连接"，要保持客户端程序的在线状态，需要不断地向服务器发起连接请求；而 TCP 连接是一种"长连接"，连接并不会主动关闭，后续的读写操作会继续使用这个连接。下面讲解 Chap5.2.7 中的 HttpClient 和 HttpServer 的实现过程，首先是 HttpClient，代码如下所示：

```java
1.  package client;
2.  import java.io.IOException;
3.  import java.io.InputStream;
4.  import java.io.OutputStream;
5.  import java.net.HttpURLConnection;
6.  import java.net.URL;
7.  import java.util.Scanner;
8.
9.
10.
11. public class HttpClient{
12.
13.     public HttpClient() {
14.         System.out.print("请输入：");
15.         Scanner scanner = new Scanner(System.in);
16.         String mes = scanner.next();
17.         String urlStr = "http://localhost/Chap5.2.7/HttpServer";
18.         URL url;
19.         try {
20.             url = new URL(urlStr);
21.             HttpURLConnection connection = (HttpURLConnection) url
22.                     .openConnection();
23.             connection.setDoOutput(true);
24.             connection.setRequestMethod("POST");
25.             OutputStream out = connection.getOutputStream();
26.             System.out.println("客户端发送的消息："+mes);
27.             out.write(mes.getBytes());
28.             InputStream in = connection.getInputStream();
29.             byte[] buffer = new byte[1024];
30.             int index = in.read(buffer);
31.             String receive = new String(buffer, 0, index);
32.             System.out.println("服务器端返回的消息："+receive);
33.             in.close();
34.             out.close();
35.         } catch (IOException e) {
36.             e.printStackTrace();
37.         }
```

```
38.
39.        }
40.
41.    public static void main(String[] args) {
42.            new HttpClient();
43.
44.                }
45.
46. }
```

HttpClient 在 client 包中，第 17 行定义了 URL 地址，第 20 行定义了 URL 对象，第 21 行得到 HttpURLConnection 对象，第 23 行将 DoOutput 标志设置为 true，指示应用程序要将数据写入 URL 连接，在第 24 行设定请求的方式是 POST 请求，第 25 行和第 28 行得到输出流和输入流，然后与服务器端交互。

接下来是 HttpServer，代码如下所示：

```
1.  package server;
2.
3.  import java.io.IOException;
4.  import javax.servlet.ServletInputStream;
5.  import javax.servlet.ServletOutputStream;
6.  import javax.servlet.http.HttpServlet;
7.  import javax.servlet.http.HttpServletRequest;
8.  import javax.servlet.http.HttpServletResponse;
9.
10.
11. public class HttpServer  extends HttpServlet{
12.
13.     public void doPost(HttpServletRequest req, HttpServletResponse resp)
14.        {
15.         try {
16.             ServletInputStream  in = req.getInputStream();
17.             ServletOutputStream out = resp.getOutputStream();
18.              int len = req.getContentLength();
19.              byte[] buffer = new byte[len];
20.              int index = in.read(buffer);
21.              String receive = new String(buffer, 0, index);
22.             System.out.println("服务器端接收到的消息："+receive);
23.             String mes="word";
24.             out.write(mes.getBytes());
25.             System.out.println("服务器端发送的消息："+mes);
26.              in.close();
27.              out.close();
28.        } catch (IOException e) {
```

```
29.                    e.printStackTrace();
30.            }
31.
32.        }
33. }
```

HttpServer 在 server 包中，HttpServer 继承了 HttpServlet，并且覆写了 doPost()方法，第 16 行和第 17 行得到了输入输出流，和客户端进行交互。客户端发送 post 请求，tomcat 将 HttpServer 放入 tomcat 中运行，并且根据请求的方式执行请求的 post 方法，然后断开连接。

5.3 通 信 协 议

5.3.1 什么是协议，为什么需要协议

协议就是一组规则。在玩游戏的时候，有游戏的规则，我们需要遵守这个规则才能进行下去。例如打麻将，首先规定了麻将的规则，是成都麻将还是重庆麻将，四个人使用相同的规则我们才能玩，如果两个人打成都麻将，两个人打重庆麻将，那么这场牌就没有办法打下去。再如插座与插头，两孔插座不能插入三头的插头，插座与插头需要接口对应才能完全吻合，这是插座与插头的规则。以上这些例子说的就是生活中潜在的协议，那么计算机之间的通信也是有协议的。落实到计算机网络，通信双方必须遵循相同的协议才能互联互通，否则无法通信。例如，我们用 QQ 客户端不能登录 MSN 的服务器。

在 5.2 节所讲的客户端和服务器端的通信也是遵守了协议的，例如，如果客户端发送 hello，则服务器端返回 word；如果客户端发送 good，则服务器端返回 thanks，这一组规则就是协议，而这些协议是我们自定义的协议。

5.3.2 如何实现协议

1．协议定义

协议有两种形式：一种是基于文本的，另一种是基于二进制的。为了方便理解，本书采用的是基于文本的协议。协议既可以是自定义的，也可以是参照格式的文本，例如 Http RFC，读者可以在 Google 上搜索 Http RFC，出现的第一条记录就是 Http 协议。5.5 节通过自定义协议使客户端与服务器端可以通信，后面将会看到。协议的定义又分请求和响应两部分，例如 5.2 节中客户端发送 hello，服务器端返回 word，就是一组请求和响应。

2．协议处理

上面讲了协议的定义。协议是由请求和响应组成的，那么如何对协议进行处理呢？首先发送方需要将消息通过指定的协议打包发送给接收方，接收方收到消息根据协议对消息进行解析，对消息的打包和解析就是对协议的处理。

5.4 Handler 机制

Handler 机制是 Android 中的一种消息异步处理机制。当应用程序启动时，Android 首先会开启一个主线程（也就是 UI 线程），主线程为管理界面中的 UI 控件，如果需要一个耗时的操作，例如通过网络连接读取数据库，如果把耗时操作放在主线程中，界面会出现假死现象。如果 5s 还没有完成的话，会收到 Android 系统的一个错误提示"强制关闭"。为了解决这个问题，需要把耗时操作放在一个子线程中，由于子线程涉及 UI 更新，所以 Android 主线程是线程不安全的，也就是说，更新 UI 只能在主线程中更新，在子线程中操作是危险的。这时就使用 Handler 对 UI 进行更新，图 5-26 为子线程通过 Handler 更新 UI（主线程）的示意图。

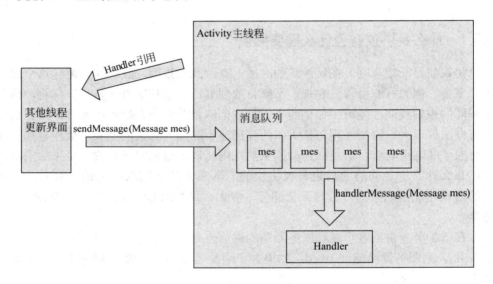

图 5-26　子线程与主线程

从图 5-26 可以看到，首先主线程将 Handler 对象的引用传递给子线程，子线程进行一些操作后要对主线程界面进行更新。此时在子线程中调用 Handler 的 sendMessage 方法，将封装到 Message 对象中数据发送到主线程的消息队列中，主线程的 Handler 从消息队列中取出消息，通过 handlerMessage 方法对取出的消息进行处理，Handler 类还有其他的一些方法，读者可以查阅 Android API 文档。Android 中对界面的更新和 Java 中的 Swing 机制有相似之处，读者可以参考《Java 核心技术卷一》的第 14 章。

5.5 联网的图书管理系统

本章学习了 Android 的 TCP 和 HTTP 编程，对 Handler 机制也做了介绍，下面通过两个实例——使用 TCP Socket 的图书管理系统和使用 HTTP 的图书管理系统讲述本章所学内容的应用。

5.5.1 定义协议

Android 客户端需要与服务器端进行通信,首先需要自定义客户端和服务器端的通信协议,协议如下所示:

插入图书
客户端:
operate:insert
content:图书
服务器端:
operate:insert
content:图书列表
result: 插入成功/插入失败,已有此图书
删除图书
客户端:
operate:delete
content:图书名称
服务器端:
operate:delete
content:图书列表
result: 删除成功/删除失败,没有此图书
修改图书
客户端:
operate:set
content:图书
服务器端:
operate:set
content:图书列表
result: 修改成功/修改失败,没有此图书
查询图书
客户端:
operate:select
content:
服务器端:
operate:select
content:图书列表
result: 查询成功/查询失败
退出

客户端：

operate:exit

content

我们自定义了增、删、改、查 4 种协议消息。每次客户端发送增加图书的请求时，就会将 operate:insert 和 content:新增加的图书打包成一个大的字符串发送到服务器端，服务器端接收到这个消息时，对此消息进行解析并响应，服务器端在响应客户端时，首先打包消息如：

operate:insert

content:图书列表

result:插入成功/插入失败，已有此图书

客户端和服务器端就通过这样的协议进行通信。

5.5.2 使用 TCP Socket 的图书管理系统

Chapter05_tcp 工程是在 Chapter03 工程上改进的，增加了 TCP 网络连接的功能，所以本节主要讲解 TCP 网络连接和更新 UI 界面。Chapter05_tcp 工程有 5 个包，分别是 control.cqupt、model.cqupt、net.cqupt、ui.cqupt、util.cqupt。与 Chapter03 相比，增加了 net.cqupt 包、util.cqupt 包并在 model.cqupt 包中增加了 Response 类，net.cqupt 包的功能是建立 TCP 网络连接，util.cqupt 包的功能是对发送和接收的数据进行打包和解析，Response 类的功能是存储服务器发送的所有信息，Chapter05_tcp 包结构如图 5-27 所示。

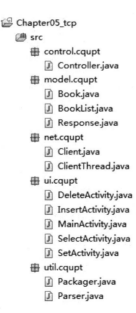

图 5-27 包结构

下面来看看 Chapter05_tcp 各个类之间的关系，如图 5-28 所示。

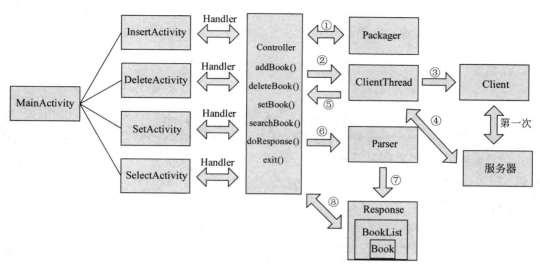

图 5-28　各个类之间的关系

下面开始按照图 5-28 标记的步骤讲解各个类。我们主要针对增加图书的流程做详细讲解，删除、修改和查询图书的实现和增加图书基本相同。**Controller** 的代码如下所示：

```
1.  package control.cqupt;
2.  import java.io.IOException;
3.  import java.io.OutputStream;
4.  import android.os.Bundle;
5.  import android.os.Handler;
6.  import android.os.Message;
7.  import net.cqupt.Client;
8.  import net.cqupt.ClientThread;
9.  import util.cqupt.Packager;
10. import util.cqupt.Parser;
11. import model.cqupt.Response;
12.
13. public class Controller {
14.     private Handler handler;
15.
16.     public Controller() {
17.
18.     }
19.
20.     public Controller(Handler handler) {
21.         this.handler = handler;
22.     }
23.
24.     public void addBook(String id,String name,String price) {
25.         Packager packager = new Packager();
```

```
26.         String message = packager.addPackage(id,name,price);
27.         new ClientThread(this, message).start();
28.
29.     }
30.
31.     public void searchBook() {
32.         Packager packager = new Packager();
33.         String message = packager.searchPackage();
34.         new ClientThread(this, message).start();
35.
36.     }
37.
38.     public void deleteBook(String name) {
39.         Packager packager = new Packager();
40.         String message=packager.deletePackage(name);
41.         new ClientThread(this, message).start();
42.     }
43.
44.     public void setBook(String id,String name,String price) {
45.         Packager packager = new Packager();
46.         String message = packager.setPackage(id,name,price);
47.         new ClientThread(this, message).start();
48.
49.     }
50.
51.     public void doResponse(String message) {
52.         Parser parser=new Parser();
53.         Response response=parser.parserResponse(message);
54.         String operate = response.getOperate();
55.         String result = response.getResult();
56.         Message msg = new Message();
57.         Bundle bundle = new Bundle();
58.         if (operate.equals("insert")) {
59.             bundle.putString("result", result);
60.             msg.setData(bundle);
61.             handler.sendMessage(msg);
62.         } else if (operate.equals("delete")) {
63.             bundle.putString("result", result);
64.             msg.setData(bundle);
65.             handler.sendMessage(msg);
66.         }else if (operate.equals("set")) {
67.             bundle.putString("result", result);
68.             msg.setData(bundle);
69.             handler.sendMessage(msg);
```

```
70.            }else if (operate.equals("select")) {
71.                bundle.putString("result", result);
72.                msg.setData(bundle);
73.                handler.sendMessage(msg);
74.            }
75.        }
76.
77.        public void exit() {
78.            Packager packager = new Packager();
79.            String message = packager.exitPackage();
80.            OutputStream out = null;
81.            try {
82.                out = Client.getSocket().getOutputStream();
83.                out.write(message.getBytes("GBK"));
84.                out.flush();
85.                out.close();
86.            } catch (IOException e) {
87.                e.printStackTrace();
88.            } finally {
89.                Client.close();
90.            }
91.
92.        }
93. }
```

Controller 在第 14 行声明了一个私有的成员变量 handler 对象，第 20～22 行通过构造数传递的参数为 handler 对象赋值，传递的 handler 对象是在 UI 界面定义的。关于增加图书功能，InsertActivity 将 handler 对象传递给 Controller 构造函数，并且调用 addBook 成员方法，Controller 类中第 24～29 行是 addBook 方法的实现，第 25、26 行打包数据并返回打包后的字符串。在第 27 行创建一个发送线程 ClientThread 并且启动线程，把打包后的数据和 Controller 类对象作为参数传递到线程中。传递 Controller 类对象的 this 指针是因为在 ClientThread 接收到服务器的返回值需要调用 Controller 类的 doResponse 方法，这些步骤都会在后面讲到。deleteBook 方法、setBook 方法、searchBook 方法用了同样的实现功能，不同的是打包方式。第 77～92 行是退出程序时执行的方法 exit，在 exit 方法中直接打包和发送信息给服务器，没有开新的线程，第 78、79 行打包数据并返回打包好的字符串，第 82～85 行得到 Socket 网络连接，并发送信息，第 84 行的 flush 方法是刷新缓存，在第 85 行关闭输出流，最后在第 89 行调用 Client 中的 close 方法关闭 Socket 连接。

下面讲解图 5-28 的第①步 util.cqupt 中的 Packager 类，Packager 类的作用是打包发送的数据。打包数据是按照自己规定的协议来打包的。Packager 的代码如下所示：

```
1.  package util.cqupt;
```

```java
2.
3.  public class Packager {
4.      public String addPackage(String id, String name, String price) {
5.          StringBuffer mes = new StringBuffer("");
6.          mes.append("operate:insert" + "\n");
7.          String content = BookPackage(id, name, price);
8.          mes.append("content:" + content + "\n");
9.          return mes.toString();
10.     }
11.
12.     public String searchPackage() {
13.         StringBuffer mes = new StringBuffer("");
14.         mes.append("operate:select" + "\n");
15.         mes.append("content:" + "\n");
16.         return mes.toString();
17.
18.     }
19.
20.     public String deletePackage(String name) {
21.         StringBuffer mes = new StringBuffer("");
22.         mes.append("operate:delete" + "\n");
23.         mes.append("content:" + name + "\n");
24.         return mes.toString();
25.     }
26.
27.     public String setPackage(String id, String name, String price) {
28.         StringBuffer mes = new StringBuffer("");
29.         mes.append("operate:set" + "\n");
30.         String content = BookPackage(id, name, price);
31.         mes.append("content:" + content + "\n");
32.         return mes.toString();
33.     }
34.
35.     public String exitPackage() {
36.         StringBuffer mes = new StringBuffer("");
37.         mes.append("operate:exit" + "\n");
38.         mes.append("content:" + "\n");
39.         return mes.toString();
40.     }
41.
42.     public String BookPackage(String id, String name, String price) {
43.         StringBuffer str = new StringBuffer("");
44.         str.append(id + "#" + name + "#" + price);
45.         return str.toString();
```

```
46.     }
47.
48. }
```

Packager 类是对数据打包的工具类，实现了增、删、改、查、退出程序和图书信息打包的功能。这里主要讲解增加图书的打包方法。在第 4~10 行的 addPackage 方法中，在第 5 行新建一个 StringBuffer 对象，第 6 行在 StringBuffer 对象中按照自定义的协议加入操作的方法，在第 7 行对整本通过 BookPackage 打包图书并返回打包后的字符串，BookPackage 方法是在第 42~46 行实现的，对图书的打包也是根据协议完成的。在 addPackage 方法的第 9 行返回打包后的数据。deletePackage 方法、setPackage 方法、searchPackage 方法和 exitPackage 方法的实现与 addPackage 方法的实现相同，不同的是打包的内容。读者可自行阅读这些方法的实现，此处不再赘述。

图 5-28 的第②步每次向服务器发送数据都新开一个 ClientThread 线程，net.cqupt 包中的 ClientThread 的代码如下：

```
1.  package net.cqupt;
2.
3.  import java.io.IOException;
4.  import java.io.InputStream;
5.  import java.io.OutputStream;
6.  import java.net.Socket;
7.
8.  import control.cqupt.Controller;
9.
10. public class ClientThread extends Thread {
11.     private InputStream in = null;
12.     private OutputStream out = null;
13.     private static final int SIZE = 1024;
14.     private String mes;
15.     private Controller controller;
16.
17.     public ClientThread(Controller controller, String mes) {
18.         Socket socket = Client.getSocket();
19.         this.controller = controller;
20.         this.mes = mes;
21.         try {
22.             out = socket.getOutputStream();
23.             in = socket.getInputStream();
24.         } catch (IOException e) {
25.             e.printStackTrace();
26.         }
27.     }
28.
```

```
29.    public void run() {
30.
31.        try {
32.            send();
33.            byte[] buffer = new byte[SIZE];
34.            int index = in.read(buffer);
35.            String message = new String(buffer, 0, index, "GBK");
36.            controller.doResponse(message);
37.        } catch (IOException e) {
38.            e.printStackTrace();
39.        }
40.
41.    }
42.
43.    public void send() {
44.        try {
45.            out.write(mes.getBytes("GBK"));
46.            out.flush();
47.        } catch (IOException e) {
48.            e.printStackTrace();
49.        }
50.    }
51. }
```

ClientThread 继承了 Thread 类，并且在 run 方法中实现了发送和接收数据的功能。ClientThread 类在第 11、12 行声明了输入输出流的，在第 13 行定义了 Buffer 数组的大小，第 14 行声明了一个私有成员变量 mes 存放 Controller 类传递过来的打包后的字符串。在第 17~27 行的构造函数中，首先通过 Client 获得 socket 连接，这是图 5-28 的第③步，Client 的代码如下：

```
1.  package net.cqupt;
2.
3.  import java.io.IOException;
4.  import java.net.Socket;
5.  import java.net.UnknownHostException;
6.
7.  public class Client {
8.      private static Socket socket;
9.
10.     private Client() {
11.         try {
12.             socket = new Socket("192.168.1.101", 8002);
13.         } catch (UnknownHostException e) {
14.             e.printStackTrace();
```

```
15.            } catch (IOException e) {
16.                e.printStackTrace();
17.            }
18.        }
19.
20.        public static Socket getSocket() {
21.            if (socket == null)
22.                new Client();
23.            return socket;
24.        }
25.
26.        public static void close() {
27.            if (socket != null)
28.                try {
29.                    socket.close();
30.                    socket = null;
31.                } catch (IOException e) {
32.                    e.printStackTrace();
33.                }
34.        }
35.
36.    }
```

Client 类和 Chapter03 中的 BookList 类一样都用了单例模式，关于单例模式在这里就不做详细讲解了。Client 类在第 10～18 行的私有构造函数中连接服务器，第 12 行 Socket 构造函数的第一个参数为服务器的 IP 地址（注意：由于用的是 Android 开发，所以不能用 localhost 代表服务器的 IP 地址），第二个参数为服务器的端口号。在 Client 中还有一个 close 方法，用于关闭 Socket 连接，并且释放 Socket 对象。

下面回到 ClientThread 类中，在 ClientThread 构造函数中获得 Socket 连接后，接着为成员变量 controller 和 mes 赋值，第 22、23 行获得 Socket 连接的输入输出管道。ClientThread 有一个 send 方法（第 43～50 行），这个方法实现了向服务器发送数据，在第 45 行通过 OutputStream 的 write 方法发送 mes 数据，在 write 方法中把 mes 通过 getBytes 方法变成 byte 数组，并且指定了编码为 GBK。在第 46 行调用 flush 方法刷新缓冲区。现在讲解 ClientThread 的核心方法 run。一旦线程启动，就开始执行 run 方法（第 29～41 行），run 方法中首先调用 send 方法发送数据，第 33～35 行实现接收服务器返回的信息，将得到的数据读入 byte 数组中，然后转换成编码格式为 GBK 的字符串，这是图 5-29 的第④步。在第 36 行调用 Controller 类中的 doResponse 方法对服务器返回的数据进行处理，这是图 5-28 的第⑤步，Controller 类中 doResponse 方法的代码如下：

```
1.  public void doResponse(String message) {
2.      Parser parser=new Parser();
3.      Response response=parser.parserResponse(message);
```

```
4.          String operate = response.getOperate();
5.          String result = response.getResult();
6.          Message msg = new Message();
7.          Bundle bundle = new Bundle();
8.          if (operate.equals("insert")) {
9.              bundle.putString("result", result);
10.             msg.setData(bundle);
11.             handler.sendMessage(msg);
12.         } else if (operate.equals("delete")) {
13.             bundle.putString("result", result);
14.             msg.setData(bundle);
15.             handler.sendMessage(msg);
16.         }else if (operate.equals("set")) {
17.             bundle.putString("result", result);
18.             msg.setData(bundle);
19.             handler.sendMessage(msg);
20.         }else if (operate.equals("select")) {
21.             bundle.putString("result", result);
22.             msg.setData(bundle);
23.             handler.sendMessage(msg);
24.         }
25.     }
```

doResponse 方法是对服务器返回的数据进行处理，第 2、3 行是对服务器返回数据的分析并将数据存储在 Response 类中，这是图 5-28 的第⑥步和第⑦步。那么首先来看一下 Parser 类的代码：

```
1.  package util.cqupt;
2.
3.  import model.cqupt.Book;
4.  import model.cqupt.BookList;
5.  import model.cqupt.Response;
6.
7.  public class Parser {
8.
9.      public Response parserResponse(String mes) {
10.         String[] message = mes.toString().split("\n");
11.         int length = message.length;
12.         String operate = message[0].substring(8, message[0].length());
13.         String result = message[length - 1].substring(7,
14.                 message[length - 1].length());
15.         Response response = Response.getResponse();
16.         response.setOperate(operate);
17.         response.setResult(result);
```

```
18.            BookList booklist = new BookList();
19.            for (int i = 1; i < message.length - 1; ++i) {
20.                String str = null;
21.                if (i == 1) {
22.                    str = message[1].substring(8, message[1].length());
23.                } else {
24.                    str = message[i];
25.                }
26.                Book book = parseBook(str);
27.                booklist.add(book);
28.            }
29.            response.setBooklist(booklist);
30.            return response;
31.        }
32.
33.        public Book parseBook(String str) {
34.            String[] mes = str.split("#");
35.            String id = mes[0];
36.            String name = mes[1];
37.            String price = mes[2];
38.            return new Book(id, name, price);
39.        }
40.
41.    }
```

Parser 类中第 9~31 行为成员方法 parserResponse。在这个方法中实现了解析数据，将数据存放到 Response 类中并且返回 Response 对象。现在详细分析 parserResponse 方法。在第 10 行过滤掉数据的换行符，将内容放于 message 数组中，在第 11 行得到 message 数组长度，在第 12~14 行根据协议提取 message 数组中的内容，第 15~29 行把在 message 数组中提取的信息放入 Response 类中保存，parseBook 方法的作用是解析图书的信息。model.cqupt 包中 Response 类的代码如下：

```
1.   package model.cqupt;
2.
3.   public class Response {
4.       private static Response response;
5.       private BookList booklist;
6.       private String result;
7.       private String operate;
8.
9.       private Response() {
10.      }
11.
12.      public static Response getResponse() {
```

```
13.            if (response == null)
14.                response = new Response();
15.            return response;
16.        }
17.
18.        public void setOperate(String operate) {
19.            this.operate = operate;
20.        }
21.
22.        public void setResult(String result) {
23.            this.result = result;
24.        }
25.
26.        public void setBooklist(BookList booklist) {
27.            this.booklist = booklist;
28.        }
29.
30.        public String getOperate() {
31.            return operate;
32.        }
33.
34.        public String getResult() {
35.            return result;
36.        }
37.
38.        public BookList getBookList() {
39.
40.            return booklist;
41.        }
42.    }
```

Response 类与 Client 类有相似之处，都只能被创建一次。Response 类有 4 个成员变量：response、operate、result 和 booklist，由名字可以知道这是服务器返回的全部信息，Response 就是一个存放返回内容的结果类。Response 定义了一个私有构造函数，防止外部调用。第 12～16 行 getResponse 方法的作用是得到 Response 对象。第 18～41 行通过 set 和 get 方法设置和得到成员服务器返回结果的信息，如 operate、result 和 booklist。

现在回到 Parser 类中的 parserResponse 方法，在 parserResponse 方法将所有解析后的结果都放入到 Response 中后，返回了 Response 对象给 Controller 类的 doResponse 方法，这是图 5-28 的第⑧步。接下来继续讲解 doResponse 方法中的剩余操作，在 doResponse 方法的第 4、5 行得到返回的方法和结果，在第 6、7 行定义一个 Message 对象和 Bundle 对象，在第 8～24 行通过判断在第 4 行得到的 operate 来执行不同的操作，例如第一个 insert 操作，在第 9 行将返回的结果存入 Bundle 中，在第 10 行将 bundle 对象放入 Message

对象里，在第 11 行，通过前面所学的 handler 用法，调用 handler 的 sendMessage 方法将数据发送到 InsertActivity 界面的消息队列中，InsertActivity 中定义的 handler 从消息队列中取出数据，通过 handlerMessage 方法对取出的数据进行处理，对界面进行更新操作。InsertActivity 定义 handler 的代码如下所示：

```
1.   private Handler handler = new Handler() {
2.         public void handleMessage(Message msg) {
3.             super.handleMessage(msg);
4.             Bundle b = msg.getData();
5.             String result = b.getString("result");
6.             buildDialog(result);
7.         }
8.   };
```

这是个匿名内部类，在第 2~7 行的 handleMessage 方法对 InsertActivity 界面更新，在第 4 行从 Message 对象中得到传递到消息队列中的数据，在第 5 行由 Bundle 的 getString 方法通过名称取得存放在 Bundle 的数据，然后通过调用 buildDialog 创建对话框并将结果显示到对话框中。在 doResponse 方法的其他判断中，都采用了同样的思想，此处不再赘述。

至此就完成了对使用 TCP Socket 的图书管理系统的讲解，下面继续讲解本节中所用服务器的实现。

5.5.3 使用 TCP Socket 的图书管理系统的服务器

本节讲述 5.4.1 节使用的服务器。首先运行 Server，运行效果如图 5-29 和图 5-30 所示。

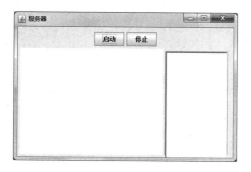
图 5-29 Server 运行效果

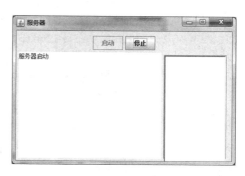
图 5-30 单击"启动"按钮

服务器端是用纯 Java 编写的，图 5-31 为 Server 工程的包结构。

ui.cqupt 包中是服务器的界面，model.cqupt 是模型，control.cqupt 包起到控制作用，控制 model.cqupt 中的 BookList 和 ui.cqupt 包的界面。util.cqupt 是工具包，对服务器接收和发送的数据打包和解析，net.cqupt 包实现 TCP 网络连接，data.cqupt 包实现 MySQL 数据库的连接。

下面将按照服务器端执行顺序进行讲解。首先是 ui.cqupt 包中的 MainClass，这个类是 Server 的界面，代码如下：

图 5-31　包结构

```
1.  package ui.cqupt;
2.
3.  import java.awt.BorderLayout;
4.  import java.awt.EventQueue;
5.  import java.awt.FlowLayout;
6.  import java.awt.List;
7.  import java.awt.event.ActionEvent;
8.  import java.awt.event.ActionListener;
9.
10. import javax.swing.JButton;
11. import javax.swing.JFrame;
12. import javax.swing.JOptionPane;
13. import javax.swing.JPanel;
14. import javax.swing.JTextArea;
15. import javax.swing.border.EmptyBorder;
16. import javax.swing.JScrollPane;
17.
18. import net.cqupt.Server;
19.
20. public class MainClass extends JFrame implements ActionListener {
21.
22.     private JTextArea textArea = new JTextArea();
23.     private JButton start = new JButton("启动");
24.     private JButton stop = new JButton("停止");
25.     private List list = new List(35);
26.     private Server server;
27.
28.     public static void main(String[] args) {
29.         EventQueue.invokeLater(new Runnable() {
30.             public void run() {
31.                 try {
32.                     MainClass frame = new MainClass();
33.                     frame.setVisible(true);
34.                 } catch (Exception e) {
35.                     e.printStackTrace();
36.                 }
37.             }
38.         });
39.     }
40.
41.     public MainClass() {
```

```java
42.        setTitle("服务器");
43.        setDefaultCloseOperation(JFrame.EXIT_ON_CLOSE);
44.        setBounds(100, 100, 450, 300);
45.        JPanel contentPane = new JPanel();
46.        JPanel right = new JPanel();
47.        contentPane.setBorder(new EmptyBorder(5, 5, 5, 5));
48.        setContentPane(contentPane);
49.        contentPane.setLayout(new BorderLayout(0, 0));
50.
51.        JPanel panel = new JPanel();
52.        contentPane.add(panel, BorderLayout.NORTH);
53.        panel.setLayout(new FlowLayout(FlowLayout.CENTER, 5, 5));
54.
55.        panel.add(start);
56.        start.addActionListener(this);
57.
58.        panel.add(stop);
59.        stop.addActionListener(this);
60.        contentPane.add(textArea, BorderLayout.CENTER);
61.        right.add(new JScrollPane(list));
62.        getContentPane().add(right, BorderLayout.EAST);
63.        list.addActionListener(new ActionListener() {
64.
65.            public void actionPerformed(ActionEvent e) {
66.                String item = list.getSelectedItem();
67.                if (JOptionPane.showConfirmDialog(null, "是否断开与当前客户端的连接") == JOptionPane.OK_OPTION) {
68.                    server.deleteThread(item);
69.                    deleteList(item);
70.                    textArea.append("客户端" + item + "连接中断" + "\n");
71.
72.                }
73.
74.            }
75.
76.        });
77.
78.    }
79.
80.    public void actionPerformed(ActionEvent e) {
81.        if (e.getSource() == start) {
82.            server = new Server(this);
83.            server.start();
84.            textArea.append("服务器启动" + "\n");
```

```
85.             start.setEnabled(false);
86.         }
87.         if (e.getSource() == stop) {
88.             if (server != null)
89.                 server.interrupt();
90.             System.exit(0);
91.         }
92.     }
93.
94.     public void addList(String client) {
95.         list.add(client);
96.     }
97.
98.     public void deleteList(String item) {
99.         list.remove(item);
100.    }
101.
102.    public void setTextArea(String str) {
103.        textArea.append(str + "\n");
104.    }
105.
106. }
```

MainClass 类用到了 Java 图形设计和事件监听的知识，下面对这段代码进行分析。MainClass 首先继承了 JFrame 和实现了 ActionListener。在第 41～78 行的构造函数中，我们构建了整个图形化界面，在第 56 行为启动按钮添加了监听器，在第 58 行为停止按钮添加了监听器。第 63～76 行是为 list 组件添加事件监听，并且实现监听器，list 组件显示所有连入的客户端，第 66 行通过 getSelectedItem 方法得到 list 上的条目名称，双击 list 上的条目就会执行第 67～72 行，生成一个判断框，当单击"确定"按钮时执行第 68～70 行，第 68 行删除此线程，在第 69 行调用在第 98～100 行定义的 deleteList 方法删除 list 组件中的这个条目。第 70 行在 Server 界面的 textArea 组件上显示客户端断开的信息。在第 80～92 行是启动和停止按钮的监听器，第 81～86 行是单击启动按钮执行的代码，单击启动按钮，首先实例化 Server 对象并且把整个 MainClass 对象传递给 Server，Server 是一个线程类，接收客户端的连接，后面将会详细讲解。第 83 行启动 Server 线程，第 84、85 行对界面操作，在 textArea 显示有客户端连接，把启动按钮设置成不可再次单击。第 87～91 行是单击退出按钮执行的代码，首先判断 Server 对象是否为 null，如果不是，则中断 Service 线程，然后退出界面；否则直接退出界面。第 94～96 行是 addList 对 list 的增加操作，每次连入一个客户端，在 Server 中调用就这个方法。第 102～104 是在 textArea 上显示信息上调用的。这就是 MainClass 的全部内容。

下面开始讲解 net.cqupt 包中的 Server 和 ServerThread。Server 类的作用是等待客户端的连接，一旦有新的客户端连接上，就会为这个客户端新建一个收发线程

ServerThread，服务器在 ServerThread 中接收客户端数据，并且对之解析、打包，最后发送回客户端。Server 类代码如下：

```
1.  package net.cqupt;
2.
3.  import java.io.IOException;
4.  import java.net.ServerSocket;
5.  import java.net.Socket;
6.  import java.util.HashMap;
7.
8.  import ui.cqupt.MainClass;
9.
10. import control.cqupt.Controller;
11.
12. public class Server extends Thread {
13.     private ServerSocket serverscoket;
14.     private HashMap<String, ServerThread> threadPool = new HashMap
            <String, ServerThread>();
15.     private Controller controller;
16.
17.     public Server(MainClass mainClass) {
18.         controller = new Controller(mainClass);
19.     }
20.
21.     public void run() {
22.         try {
23.             serverscoket = new ServerSocket(8002);
24.             while (!Server.interrupted()) {
25.                 Socket socket = serverscoket.accept();
26.                 String name = socket.getLocalAddress().
                        getHostAddress() + "::"
27.                         + socket.getPort();
28.                 controller.addClientView(name);
29.                 ServerThread st = new ServerThread(name, socket,
                        this);
30.                 threadPool.put(name, st);
31.                 st.start();
32.             }
33.         } catch (IOException e) {
34.             e.printStackTrace();
35.         } finally {
36.             close();
37.         }
38.
```

```
39.        }
40.
41.        public void deleteThread(String name) {
42.            ServerThread thread = threadPool.get(name);
43.            thread.interrupt();
44.            threadPool.remove(name);
45.        }
46.
47.        public Controller getController() {
48.            return controller;
49.        }
50.
51.        public void close() {
52.            if (serverscoket != null) {
53.                try {
54.                    serverscoket.close();
55.                } catch (IOException e) {
56.                    e.printStackTrace();
57.                }
58.            }
59.        }
60. }
```

Server 类继承了 Thread 类，在第 17～19 行的构造函数中，接收 MainClass 对象作为参数，在第 18 行实例化 control 对象并且把 MainClass 对象作为参数传递给 Controller。21～39 行是线程执行的 run 方法，第 23 行使用 ServerSocket(int port)实例化一个 ServerSocket 对象，port 参数传递端口号，这个端口就是服务器监听连接请求的端口，如果在这时出现错误，将抛出 IOException 异常对象并且执行在第 51～59 行定义的 close 方法关闭 ServerSocket 连接，否则将实例化 ServerSocket 对象并开始准备接收连接请求。接下来中第 24～32 行服务程序进入无限循环中，无限循环从第 25 行调用 ServerSocket 的 accept 方法开始，在调用开始后 accept 方法将导致调用线程阻塞直到连接建立。在建立连接后 accept 返回一个最近创建的 Socket 对象，该 Socket 对象绑定了客户程序的 IP 地址或端口号。在第 26 行得到客户端的 IP 地址和端口号，在第 28 行调用 Controller 类中的 addClientView 方法在界面显示客户端连接上的信息，Controller 类将在后面介绍。第 29 行为客户端新开一个线程，将客户端的名称、Socket 对象和 Server 对象作为参数传递给 ServerThread 线程类。在第 30 行将这个 ServerThread 线程添加到在第 14 行定义的 threadPool 成员变量中保存，然后调用 start 运行线程，threadPool 的类型是 HashMap，以客户端名称作为 key，ServerThread 对象作为 value 存储。在第 41～45 行 deleteThread 方法是删除在 threadPool 中的线程并且在第 43 行调用线程的 interrupt 方法终止线程。第 47～49 行的 getController 是得到 control 对象，这个方法将在 ServerThread 中用到。现在就完成了对 Server 的讲解。

下面开始讲解 ServerThread 类，这是个接收和发送数据的类，和 5.4 节所讲的 ClientThread 有相同的功能。ServerThread 代码如下所示：

```java
1.  package net.cqupt;
2.
3.  import java.io.IOException;
4.  import java.io.InputStream;
5.  import java.io.OutputStream;
6.  import java.net.Socket;
7.
8.  import util.cqupt.Parser;
9.
10. import control.cqupt.Controller;
11.
12. public class ServerThread extends Thread {
13.     private Socket socket;
14.     private String name;
15.     private InputStream in;
16.     private OutputStream out;
17.     private Server server;
18.
19.     private static final int SIZE = 1024;
20.
21.     public ServerThread(String name, Socket socket, Server server) {
22.         this.name = name;
23.         this.server = server;
24.         this.socket = socket;
25.         try {
26.             in = socket.getInputStream();
27.             out = socket.getOutputStream();
28.         } catch (IOException e) {
29.             e.printStackTrace();
30.         }
31.     }
32.
33.     public void run() {
34.         try {
35.             while (!this.isInterrupted()) {
36.                 byte[] buffer = new byte[SIZE];
37.                 int index = in.read(buffer);
38.                 String message = new String(buffer, 0, index, "GBK");
39.                 Parser parser = new Parser();
40.                 String operate = parser.getOperate(message);
41.                 Controller controller = server.getController();
```

```
42.              if (operate.equals("exit")) {
43.                  server.deleteThread(name);
44.                  controller.deleteClientView(name);
45.                  break;
46.              }
47.              String response = controller.doResponse(message);
48.              send(response);
49.          }
50.
51.      } catch (IOException e) {
52.          e.printStackTrace();
53.
54.      } finally {
55.          close();
56.      }
57.  }
58.
59.  public void send(String mes) {
60.      try {
61.          out.write(mes.getBytes("GBK"));
62.          out.flush();
63.      } catch (IOException e) {
64.          e.printStackTrace();
65.      }
66.
67.  }
68.
69.  private void close() {
70.      try {
71.          if (in != null)
72.              in.close();
73.          if (out != null)
74.              out.close();
75.          if (socket != null)
76.              socket.close();
77.      } catch (IOException e) {
78.          e.printStackTrace();
79.      }
80.  }
81.
82. }
```

ServerThread 是一个继承了 Thread 的线程类，在 ServerThread 声明了 5 个私有成员变量：name 代表客户端的名称，socket 代表客户端，server 是 Server 对象的引用，in 和

out 是输入输出流。在第 21~31 行的构造函数里为各个成员变量赋值,并且实例化 in 和 out 流,通过 socket 的 getInputStream 和 getOutputStream 的客户端与服务器之间的通道。第 33~57 行的 run 方法实现了服务器与客户端的交互功能,首先在第 35 行判断线程是否被中断,如果没有中断,则在这个 while 循环里与客户端进行交互,在第 36 行定义一个 byte 数据,长度为 SIZE,在第 19 行定义了常量 SIZE 并复制为 1024。在第 37 行通过 InputStream 的 read 方法将客户端发来的数据读入到 byte 数组中,值得注意的是,read 方法是阻塞的,也就是说,如果客户端没有数据发送到服务器,那么程序就会阻塞在第 37 行,直到接收到了数据,才继续执行下面的代码。在第 38 行将读到的数据以 GBK 的形式转换成字符串。然后在第 39、40 行对转换后的字符串解析,并且返回解析后客户端的请求操作,在第 42~46 行是判断客户端的请求操作是否为退出,如果是退出操作,则执行第 43~45 行,第 43 行在 Server 对象中的 threadPool 中删除此线程,第 54 行在 MainClass 界面上删除此客户端信息,跳出循环执行 finally 中的 close 方法关闭流,close 方法在第 69~80 行定义,作用是关闭输入输出流和 socket 连接。如果客户端发送的是其他的请求操作,则执行第 47、48 行,第 47 行调用 controller 对象的 doResponse 方法,返回打包好的服务器响应数据,在第 48 行调用在第 59~67 行定义的 send 方法发送数据到客户端。send 方法里实现了服务器的发送功能,在第 61 行通过 write 方法以 Bytes 字节,编码方式为 GBK 发送到客户端,第 62 行刷新缓冲区。

下面讲解控制类 Controller,代码如下:

```
1.  package control.cqupt;
2.
3.  import ui.cqupt.MainClass;
4.  import util.cqupt.Packager;
5.  import util.cqupt.Parser;
6.  import model.cqupt.Book;
7.  import model.cqupt.BookList;
8.
9.  public class Controller {
10.     private MainClass mainclass;
11.     private String result;
12.
13.     public Controller(MainClass mainclass) {
14.         this.mainclass = mainclass;
15.     }
16.
17.     public String doResponse(String request) {
18.         Parser parser = new Parser();
19.         String operate = parser.getOperate(request);
20.         String content = parser.getContent(request);
21.         String response = null;
22.         if (operate.equals("insert")) {
```

```
23.            addBook(content);
24.            Packager packager = new Packager();
25.            response = packager.addPackage(operate, result);
26.        } else if (operate.equals("delete")) {
27.            deleteBook(content);
28.            Packager packager = new Packager();
29.            response = packager.deletePackage(operate, result);
30.        } else if (operate.equals("set")) {
31.            setBook(content);
32.            Packager packager = new Packager();
33.            response = packager.setPackage(operate, result);
34.        } else if (operate.equals("select")) {
35.            getBookList();
36.            Packager packager = new Packager();
37.            response = packager.searchPackage(operate, result);
38.        }
39.        return response.toString();
40.    }
41.
42.    public void addBook(String content) {
43.        Parser parser = new Parser();
44.        Book book = parser.parseBook(content);
45.        BookList booklist = new BookList();
46.        if (booklist.insert(book)) {
47.            result = "插入成功";
48.        } else {
49.            result = "插入失败,已有此图书";
50.        }
51.    }
52.
53.    public void deleteBook(String content) {
54.        BookList booklist = new BookList();
55.        if (booklist.delete(content)) {
56.            result = "删除成功";
57.        } else {
58.            result = "删除失败,没有此图书";
59.        }
60.    }
61.
62.    public void setBook(String content) {
63.        BookList booklist = new BookList();
64.        Parser parser = new Parser();
65.        Book book = parser.parseBook(content);
66.        if (booklist.set(book)) {
```

```
67.            result = "修改成功";
68.          } else {
69.            result = "修改失败,没有此图书";
70.          }
71.        }
72.
73.        public void getBookList() {
74.          BookList booklist = new BookList();
75.          if (booklist != null) {
76.            result = "查询成功 ";
77.          } else {
78.            result = "查询失败";
79.          }
80.
81.        }
82.
83.        public void deleteClientView(String name) {
84.          mainclass.setTextArea("客户端" + name + "退出");
85.          mainclass.deleteList(name);
86.        }
87.
88.        public void addClientView(String name) {
89.          mainclass.addList(name);
90.          mainclass.setTextArea(name + "客户端连接上");
91.        }
92.
93.      }
```

Controller 类实现的功能是操作 BookList 类和 MainClass 类更改界面。首先在第 13~15 行的构造函数里为 MainClass 对象赋值,MainClass 对象在第 83~86 行的 deleteClientView 方法中对 MainClass 界面的 TextArea 和 List 对象操作,实现删除客户端。MainClass 对象在第 88~91 行的 addClientView 方法中同样对 MainClass 界面的 TextArea 和 List 对象操作,实现的是增加客户端的操作。

下面讲解 Controller 类在 ServerThread 中调用的 doResponse 方法,第 17~40 行自定义了 doResponse 方法,实现对服务器端收到的消息的解析,根据客户端的请求操作 BookList,并且将响应的信息打包返回。第 18~20 行是用工具类 Parser 解析消息,并且取得解析后的请求和内容。第 22~38 行根据解析后的请求执行相应的增、删、改、查操作。这里选取增加操作进行讲解,当请求操作为 insert 时,执行第 23~25 行,第 23 行调用 addBook 方法,并把内容传递到 addBook 方法中,addBook 方法是在第 42~51 行定义的,第 43、44 行解析内容并返回 book 对象,在第 45 行创建 BookList 对象,在第 46 行调用 BookList 对象中的 insert 方法插入图书,根据成功与否,对 result 变量赋值。在 doResponse 方法的第 24、25 行完成对响应信息的打包,最后在第 39 行将打包后的数据转换成的字符串并返回。删、改、查的操作流程和增加操作是一样的，此处不再赘述。

下面讲解 util.cqupt 包中的 Parser 类和 Packager 类，这两个类是工具类，实现对数据的解析和打包功能。

```
1.  package util.cqupt;
2.
3.  import model.cqupt.Book;
4.
5.  public class Parser {
6.
7.      public String getOperate(String request) {
8.          String[] message = request.split("\n");
9.          String operate = message[0].substring(8, message[0].length());
10.         return operate;
11.     }
12.
13.     public String getContent(String request) {
14.         String[] message = request.split("\n");
15.         String content = message[1].substring(8, message[1].length());
16.         return content;
17.     }
18.
19.     public Book parseBook(String str) {
20.         String[] mes = str.split("#");
21.         String id = mes[0];
22.         String name = mes[1];
23.         String price = mes[2];
24.         return new Book(id, name, price);
25.     }
26. }
```

Parser 类中实现了对客户端发送过来的请求操作以及内容和图书信息的解析。第 7~11 行 getOperate 方法是对请求操作的解析，通过 split 函数过滤掉换行符；第 13~17 行 getContent 是对内容的解析，同样是用 split 过滤掉换行符；第 19~24 行 parseBook 是对图书信息的解析。这 3 个方法解析数据都是根据自定义的协议进行解析的。

Packager 类是对服务器响应信息按照协议进行打包的工具类。Packager 类的代码如下：

```
1.  package util.cqupt;
2.
3.  import model.cqupt.Book;
4.  import model.cqupt.BookList;
5.
6.  public class Packager {
```

```java
7.      public String addPackage(String operate, String result) {
8.          BookList booklist = new BookList();
9.          StringBuffer mes = new StringBuffer("");
10.         String books = booklist.getBookList();
11.         mes.append("operate:" + operate + "\n");
12.         mes.append("content:" + books + "\n");
13.         mes.append("result:" + result + "\n");
14.
15.         return mes.toString();
16.     }
17.
18.     public String searchPackage(String operate, String result) {
19.         BookList booklist = new BookList();
20.         StringBuffer mes = new StringBuffer("");
21.         String books = booklist.getBookList();
22.         mes.append("operate:" + operate + "\n");
23.         mes.append("content:" + books + "\n");
24.         mes.append("result:" + result + "\n");
25.         System.out.println(mes.toString());
26.         return mes.toString();
27.     }
28.
29.     public String deletePackage(String operate, String result) {
30.         BookList booklist = new BookList();
31.         StringBuffer mes = new StringBuffer("");
32.         String books = booklist.getBookList();
33.         mes.append("operate:" + operate + "\n");
34.         mes.append("content:" + books + "\n");
35.         mes.append("result:" + result + "\n");
36.
37.         return mes.toString();
38.     }
39.
40.     public String setPackage(String operate, String result) {
41.         BookList booklist = new BookList();
42.         StringBuffer mes = new StringBuffer("");
43.         String books = booklist.getBookList();
44.         mes.append("operate:" + operate + "\n");
45.         mes.append("content:" + books + "\n");
46.         mes.append("result:" + result + "\n");
47.
48.         return mes.toString();
49.     }
50.
```

```
51.    public String BookPackage(Book book) {
52.        StringBuffer str = new StringBuffer("");
53.        str.append(book.getId() + "#" + book.getName() + "#" +
           book.getPrice());
54.        return str.toString();
55.    }
56.
57. }
```

Packager 类实现了对图书信息的增、删、改、查以及打包功能。Packager 打包方法是根据我们定义的协议打包的。读者可以根据定义的协议来学习代码，这里简单地介绍一下增加操作对应的打包功能。在第 7～16 行定义了 addPackage 方法，第 8 行实例化 BookList 对象，第 10 行通过 getBookList 方法以字符串的方式返回所有的图书，接下来将服务器响应的信息放到一个大的字符串里，并返回这个字符串。删、改、查的方法类似，此处不再赘述。

接下来去看看 model 包中的 BookList 类。BookList 类存储图书并且对 MySQL 数据库进行访问，BookList 的代码如下：

```
1.  package model.cqupt;
2.
3.  import java.sql.Connection;
4.  import java.sql.ResultSet;
5.  import java.sql.SQLException;
6.  import java.sql.Statement;
7.  import java.util.ArrayList;
8.
9.  import util.cqupt.Packager;
10.
11. import data.cqupt.DBconnection;
12.
13.
14. public class BookList extends ArrayList<Book> {
15.
16.
17.     private static final long serialVersionUID = 1L;
18.
19.     public BookList() {
20.         super();
21.         String sql = "SELECT id,name,price FROM books";
22.         DBconnection d = new DBconnection();
23.         Connection con = null;
24.         Statement sta = null;
25.         ResultSet re = null;
```

```
26.        try {
27.            con = d.getConnect();
28.            sta = con.createStatement();
29.            re = sta.executeQuery(sql);
30.            while (re.next()) {
31.                add(new Book(re.getString(1), re.getString(2),
                        re.getString(3)));
32.            }
33.        } catch (SQLException e) {
34.            e.printStackTrace();
35.        } finally {
36.            d.close(con, sta, re);
37.        }
38.    }
39.
40.    public boolean insert(Book book) {
41.        if (checkId(book.getId())) {
42.            String id = book.getId();
43.            String name = book.getName();
44.            String price = book.getPrice();
45.            add(book);
46.            String sql = "INSERT INTO books(id,name,price)" +
                    "VALUES('" + id
47.                    + "','" + name + "','" + price + "')";
48.            DBconnection d = new DBconnection();
49.            Connection con = null;
50.            Statement sta = null;
51.            try {
52.                con = d.getConnect();
53.                sta = con.createStatement();
54.                sta.executeUpdate(sql);
55.            } catch (SQLException e) {
56.                e.printStackTrace();
57.            } finally {
58.                d.close(con, sta);
59.            }
60.            return true;
61.        } else {
62.            return false;
63.        }
64.    }
65.
66.    public boolean delete(String name) {
67.        if (checkName(name)) {
```

```
68.            String sql = "DELETE FROM books WHERE name='" + name + "'";
69.            DBconnection d = new DBconnection();
70.            Connection con = null;
71.            Statement sta = null;
72.            try {
73.                con = d.getConnect();
74.                sta = con.createStatement();
75.                sta.executeUpdate(sql);
76.            } catch (SQLException e) {
77.                e.printStackTrace();
78.            } finally {
79.                d.close(con, sta);
80.            }
81.            return true;
82.        } else {
83.
84.            return false;
85.        }
86.    }
87.
88.    public boolean set(Book book) {
89.        if (!checkId(book.getId())) {
90.            String id = book.getId();
91.            String name = book.getName();
92.            String price = book.getPrice();
93.            int index = getIndex(id);
94.            set(index, book);
95.            String sql = "UPDATE books SET name='" + name + "'," +
                    "price='"
96.                    + price + "' WHERE id='" + id + "'";
97.            DBconnection d = new DBconnection();
98.            Connection con = null;
99.            Statement sta = null;
100.            try {
101.                con = d.getConnect();
102.                sta = con.createStatement();
103.                sta.executeUpdate(sql);
104.            } catch (SQLException e) {
105.                e.printStackTrace();
106.            } finally {
107.                d.close(con, sta);
108.            }
109.            return true;
110.        } else {
```

```
111.            return false;
112.        }
113.    }
114.
115.    public String getBookList() {
116.        StringBuffer books = new StringBuffer("");
117.        Packager packager = new Packager();
118.        for (int i = 0; i < this.size(); ++i) {
119.            if (i == this.size() - 1) {
120.                String s = packager.BookPackage(this.get(i));
121.                books.append(s);
122.            } else {
123.                Book book = this.get(i);
124.                String str = packager.BookPackage(book);
125.                books.append(str + "\n");
126.            }
127.        }
128.        return books.toString();
129.    }
130.
131.    public boolean checkId(String id) {
132.        for (int i = 0; i < this.size(); ++i) {
133.            Book book = this.get(i);
134.            String bookid = book.getId();
135.            if (bookid.equals(id))
136.                return false;
137.        }
138.        return true;
139.    }
140.
141.    public boolean checkName(String name) {
142.        for (int i = 0; i < this.size(); ++i) {
143.            Book book = this.get(i);
144.            if (book.getName().equals(name)) {
145.                remove(i);
146.                return true;
147.            }
148.        }
149.        return false;
150.    }
151.
152.    public int getIndex(String bookid) {
153.        int i = 0;
154.        for (; i < this.size(); ++i) {
```

```
155.                Book book = this.get(i);
156.                String id = book.getId();
157.                if (id.equals(bookid)) {
158.                    break;
159.                }
160.            }
161.            return i;
162.        }
163.
164.    }
```

BookList 类继承了 ArrayList，所以可以当作 ArrayList 来使用。在第 19～38 行 BookList 的构造方法中，通过访问 MySQL 数据库，将数据库中的数据存入 BookList 对象中，相当于查询数据库的功能。第 21 行是查询数据库的 SQL 语句，第 22 行实例化 DBconnection 对象，在第 23 行声明数据库连接，在第 27 行得到连接。第 24 行是声明数据库操作，并在第 28 行通过 createStatement 方法得到 Statement 对象。第 25 行声明结果集 ResultSet，在第 29 行通过 executeQuery 执行第 21 行的 SQL 语句得到 ResultSet 对象。在第 30～32 行通过 ResultSet 对象的 next 方法遍历数据库，在第 31 行将得到的数据存放到 BookList 对象中，完成对数据库的查询功能，在构造方法里查询数据库是为了使数据库和 BookList 对象同步。在 BookList 类中主要有 4 个方法对 MySQL 数据库进行操作：增加（insert）、删除（delete）、修改（set）和查询（getBookList），这些方法的实现和构造方法差不多，读者可自行学习。

最后来看一看 data.cqupt 包中的 DBconnection 类，这个类完成了 Java 连接 MySQL 数据库的功能，代码如下所示：

```
1.  package data.cqupt;
2.  import java.sql.Connection;
3.  import java.sql.DriverManager;
4.  import java.sql.ResultSet;
5.  import java.sql.SQLException;
6.  import java.sql.Statement;
7.
8.  public class DBconnection {
9.      public static final String DBURL = "jdbc:mysql://localhost:3306/books";
10.     public static final String DBUSER = "root";
11.     public static final String DBPASS = "1111";
12.     public static final String DBRIVER = "org.gjt.mm.mysql.Driver";
13.
14.     static {
15.         try {
16.             Class.forName(DBRIVER);
17.         } catch (ClassNotFoundException e) {
```

```
18.            e.printStackTrace();
19.        }
20.    }
21.
22.    public Connection getConnect() throws SQLException {
23.        return DriverManager.getConnection(DBURL, DBUSER, DBPASS);
24.    }
25.
26.    public void close(Connection con, Statement sta, ResultSet re) {
27.        try {
28.            re.close();
29.            if (con != null && re != null) {
30.                close(con, sta);
31.            }
32.        } catch (SQLException e) {
33.            e.printStackTrace();
34.        }
35.
36.    }
37.
38.    public void close(Connection con, Statement sta) {
39.        if (con != null && sta != null) {
40.            try {
41.                sta.close();
42.                con.close();
43.
44.            } catch (SQLException e) {
45.                e.printStackTrace();
46.            }
47.        }
48.    }
49. }
```

DBconnection 完成与 MySQL 数据库的连接，首先第 9 行定义 MySQL 数据库驱动程序，第 10 行定义 MySQL 数据库连接的用户名，第 11 行定义 MySQL 数据库连接的密码，第 12 行定义 MySQL 数据库的连接地址。在第 14~20 行是一个静态代码块，静态代码块只在第一次创建 DBconnection 对象的时候执行一次，此处的作用是通过 Class.forName 加载驱动。第 22~24 行代码是连接数据库并且返回 Connection 对象的引用。第 26~28 行是两个重载的 close 方法，作用是关闭操作，按照先打开后关闭的顺序执行关闭操作。

这里完成了对 TCP 的服务器端的讲解。5.5.4 节将介绍 HTTP 的应用。

5.5.4 使用 HTTP 的图书管理系统

由本章学习的网络编程知识，可知 HTTP 的底层就是 TCP，所以本节采用 HTTP 实现的 Chapter05_http 是在 Chapter05_tcp 的 net.cqupt 包上做了少量修改。Chapter05_http 的包结构如图 5-32 所示。

可以看到，net.cqupt 包中少了 Client 类，只有 ClientThread 类，ClientThread 的代码如下所示：

图 5-32 Chapter05_http 包结构

```
1.  package net.cqupt;
2.
3.  import java.io.IOException;
4.  import java.io.InputStream;
5.  import java.io.OutputStream;
6.  import java.net.HttpURLConnection;
7.  import java.net.URL;
8.
9.  import control.cqupt.Controller;
10.
11. public class ClientThread extends Thread {
12.     private InputStream in;
13.     private OutputStream out;
14.     private static final int SIZE = 1024;
15.     private String mes;
16.     private Controller controller;
17.
18.     public ClientThread(Controller controller, String mes) {
19.         this.controller = controller;
20.         this.mes = mes;
21.
22.     }
23.
24.     public void run() {
25.
26.         String urlStr = "http://192.168.1.100:8080/FiveHttpServer/server";
27.         URL url;
28.         try {
29.             url = new URL(urlStr);
30.             HttpURLConnection connection = (HttpURLConnection) url
31.                     .openConnection();
32.             connection.setDoOutput(true);
33.             connection.setRequestMethod("POST");
```

```
34.            out = connection.getOutputStream();
35.            send();
36.            in = connection.getInputStream();
37.            byte[] buffer = new byte[SIZE];
38.            int index = in.read(buffer);
39.            String message = new String(buffer, 0, index, "GBK");
40.            controller.doResponse(message);
41.        } catch (IOException e) {
42.            e.printStackTrace();
43.        }
44.
45.    }
46.
47.    public void send() {
48.        try {
49.            out.write(mes.getBytes("GBK"));
50.            out.flush();
51.            out.close();
52.        } catch (IOException e) {
53.            e.printStackTrace();
54.        }
55.    }
56. }
```

与 Chapter05_tcp 的 ClientThread 相比，这里对 run 方法做了修改，Chapter05_tcp 采用的是长连接，而此处是采用短连接。第 26 行定义了服务器的地址（注意：不能为 localhost），第 27 行声明了 URL 对象并在第 29 行实例化，第 30、31 行打开 HTTP 连接，第 32 行设置是否向 httpUrlConnection 输出，因为在第 33 行设置定了请求方法为 POST，参数要放在 HTTP 正文内，因此需要设为 true，默认情况下是 false。然后客户端就开始发送数据了，注意在 send 方法中，每次发送了数据都执行了第 51 行关闭输出流的操作。这就是 HTTP 图书管理系统，下面讲解 HTTP 图书管理系统的服务器端代码。

5.5.5 使用 HTTP 的图书管理系统的服务器

与 TCP 图书管理系统的服务器相比较，HTTP 的服务器少了界面，并使用了 tomcat 发布。Server_http 的包结构如图 5-33 所示。

这是一个 tomcat 工程，服务器通过 tomact 部署 Web 应用程序。Server_http 只在 TCP 图书管理系统的服务器上修改了 net.cqupt 包，Server 类代码如下所示：

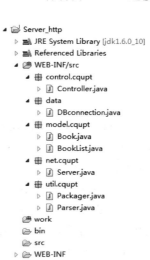

图 5-33　Server_http 包结构

```
1.  package net.cqupt;
2.
3.  import java.io.IOException;
4.  import control.cqupt.Controller;
5.  import javax.servlet.ServletInputStream;
6.  import javax.servlet.ServletOutputStream;
7.  import javax.servlet.http.HttpServlet;
8.  import javax.servlet.http.HttpServletRequest;
9.  import javax.servlet.http.HttpServletResponse;
10.
11. public class Server extends HttpServlet {
12.
13.     public void doPost(HttpServletRequest req, HttpServletResponse resp) {
14.         ServletInputStream in;
15.         ServletOutputStream out;
16.         try {
17.             in = req.getInputStream();
18.             out = resp.getOutputStream();
19.             int len = req.getContentLength();
20.             byte[] buffer = new byte[len];
21.             int index = in.read(buffer);
22.             String message = new String(buffer, 0, index, "GBK");
23.             Controller controller = new Controller();
24.             String response = controller.doResponse(message);
25.             out.write(response.getBytes("GBK"));
26.             in.close();
27.             out.close();
28.         } catch (IOException e) {
29.             e.printStackTrace();
30.         }
31.     }
32.
33. }
```

Server 继承了 HttpServlet 抽象类，并且覆盖了它的 doPost 方法。每当 Servlet 容器接收到客户端的请求，Servlet 容器就会解析 Web 客户的 HTTP 请求，然后创建一个 HttpRequest 对象，在这个对象中封装 HTTP 请求信息，创建一个 HttpResponse 对象。由于客户端是 post 请求，所以调用 HttpServlet 的 doPost 方法，把 HttpRequest 和 HttpResponse 对象作为 doPost 方法的参数传给 HttpServlet 对象。HttpServlet 调用 HttpRequest 的有关方法，获取 HTTP 请求信息，调用 HttpResponse 的有关方法，生成响应数据并传递给客户端。

第 14、15 行声明了 HTTP 的输入输出流，并在第 17、18 行实例化。其他操作和 TCP 图书管理系统的服务器端执行一样，此处不再赘述。值得注意的是，每次发送完数据都会执行第 26、27 行的 close 方法关闭输入输出流，因为 HTTP 是短连接的。

第 6 章　多　媒　体

随着世界的发展、科学的进步，多媒体技术促进了计算机科学及其相关学科的发展和融合，开拓了计算机在国民经济各个领域的应用。例如，多媒体教室、可视电话、电子杂志等。这些都对社会、经济产生了重大的影响。如今，多媒体已经在人们日常生活中深深地扎下了根，成为生活中不可或缺的一部分。本章将学习音频的播放和视频的播放，并简单介绍 MediaPlayer。通过本章的学习，帮助大家了解多媒体的有关知识，为进一步学习后续章节做好准备。

6.1　MediaPlayer

首先介绍一下 MediaPlayer。它可以用来播放音频和视频文件。虽然除了 MediaPlayer 类，SoundPool 和 JetPlayer 类也可用来播放音频文件，但这里只介绍 MediaPlayer，因为 MediaPlayer 类处于 Android media 包的核心位置。

先来看一下 MediaPlayer 类的一些常见方法，后面要讲的音频播放和视频播放会用到其中的一些方法，如表 6-1 所示。

表 6-1　MediaPlayer 类的常见方法

方　　法	描　　述
stop()	无返回值，停止播放
start()	无返回值，开始播放
setVolume(float leftVolume, float rightVolume)	无返回值，设置音量
setScreenOnWhilePlaying(boolean screenOn)	无返回值，设置是否使用 SurfaceHolder 显示
setLooping(boolean looping)	无返回值，设置是否循环播放
setDisplay(SurfaceHolder sh)	无返回值，设置用 SurfaceHolder 来显示多媒体
setDataSource(Context context, Uri uri)	无返回值，根据 Uri 设置多媒体数据来源
setDataSource(FileDescriptor fd)	无返回值，根据 FileDescriptor 设置多媒体数据来源
setDataSource(FileDescriptor fd, long offset, long length)	无返回值，根据 FileDescriptor 设置多媒体数据来源，设置最大长度
setDataSource(String path)	无返回值，根据路径设置多媒体数据来源
setAudioStreamType(int streamtype)	无返回值，指定流媒体的类型
seekTo(int msec)	无返回值，指定播放的位置（以毫秒为单位的时间）
reset()	无返回值，重置 MediaPlayer 对象
release()	无返回值，释放 MediaPlayer 对象
prepareAsync()	无返回值，准备异步

续表

方　　法	描　　述
prepare()	无返回值，准备同步
pause()	无返回值，暂停
isPlaying()	返回 boolean，是否正在播放
isLooping()	返回 boolean，是否循环播放
getVideoWidth()	返回 Int，得到视频的宽度
getVideoHeight()	返回 Int，得到视频的高度
getDuration()	返回 Int，得到文件的时间
getCurrentPosition()	返回 Int，得到当前播放位置
create(Context context, Uri uri, SurfaceHolder holder)	静态方法，通过 Uri 和指定 SurfaceHolder 创建一个多媒体播放器
create(Context context, int resid)	静态方法，通过资源 ID 创建一个多媒体播放器
create(Context context, Uri uri)	静态方法，通过 Uri 创建一个多媒体播放器

在前面的学习中，我们知道 Activity 是有生命周期的，MediaPlayer 也是有生命周期的。如图 6-1 所示的 Android MediaPlayer 状态转换图表征了它的生命周期，搞清楚这个图可以帮助我们在使用 Android MediaPlayer 时考虑更周全，写出的代码也更条理清晰。

这张状态转换图清晰地描述了 MediaPlayer 的各个状态，也列举了主要方法的调用时序，每种方法只能在一些特定的状态下使用。如果使用时 MediaPlayer 的状态不正确，则会引发 IllegalStateException 异常。

Idle 状态：当使用 new()方法创建一个 MediaPlayer 对象或者调用了其 reset()方法时，该 MediaPlayer 对象处于 Idle 状态。这两种方法的一个重要差别就是：如果在这个状态下调用了 getDuration()等方法（相当于调用时机不正确），通过 reset()方法进入 Idle 状态的话会触发 OnErrorListener.onError()，并且 MediaPlayer 会进入 Error 状态；如果是新创建的 MediaPlayer 对象，则并不会触发 onError()，也不会进入 Error 状态。

End 状态：通过 release()方法可以进入 End 状态，只要 MediaPlayer 对象不再被使用，就应当尽快将其通过 release()方法释放掉，以释放相关的软硬件组件资源，这其中有些资源是只有一份的（相当于临界资源）。如果 MediaPlayer 对象进入了 End 状态，则不会再进入任何其他状态了。

Initialized 状态：这个状态比较简单，MediaPlayer 调用 setDataSource()方法就进入 Initialized 状态，表示此时要播放的文件已经设置好了。

Prepared 状态：初始化完成之后还需要通过调用 prepare()或 prepareAsync()方法，这两个方法一个是同步的一个是异步的，只有进入 Prepared 状态，才表明 MediaPlayer 到目前为止都没有错误，可以进行文件播放。

Preparing 状态：这个状态比较好理解，主要是和 prepareAsync()配合，如果异步准备完成，会触发 OnPreparedListener.onPrepared()，进而进入 Prepared 状态。

Started 状态：显然，MediaPlayer 一旦准备好，就可以调用 start()方法，这样 MediaPlayer 就处于 Started 状态，这表明 MediaPlayer 正在播放文件过程中。可以使用 isPlaying()测试 MediaPlayer 是否处于 Started 状态。如果播放完毕，而又设置了循环播放，则 MediaPlayer 仍然会处于 Started 状态；类似地，如果在该状态下 MediaPlayer 调用了 seekTo()或者 start()方法均可以让 MediaPlayer 停留在 Started 状态。

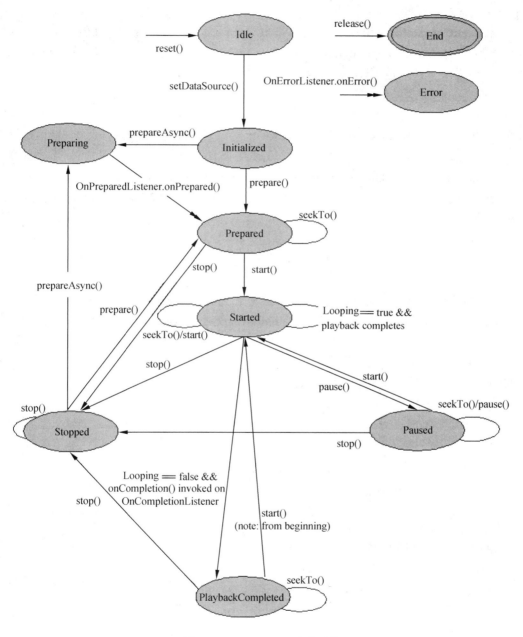

图 6-1　MediaPlayer 生命周期图

Paused 状态：Started 状态下 MediaPlayer 调用 pause()方法可以暂停 MediaPlayer，从而进入 Paused 状态，MediaPlayer 暂停后再次调用 start()则可以继续 MediaPlayer 的播放，转到 Started 状态，暂停状态时可以调用 seekTo()方法，这是不会改变状态的。

Stop 状态：Started 或者 Paused 状态下均可调用 stop()停止 MediaPlayer，而处于 Stop 状态的 MediaPlayer 要想重新播放，需要通过 prepareAsync()和 prepare()回到先前的 Prepared 状态重新开始才可以。

PlaybackCompleted 状态：文件正常播放完毕，而又没有设置循环播放的话就进入该状态，并会触发 OnCompletionListener 的 onCompletion()方法。此时可以调用 start()方法重新从头播放文件，也可以 stop()停止 MediaPlayer，还可以 seekTo()来重新定位播放位置。

Error 状态：如果由于某种原因 MediaPlayer 出现了错误，会触发 OnErrorListener.onError()事件，此时 MediaPlayer 即进入 Error 状态，及时捕捉并妥善处理这些错误是很重要的，可以帮助我们及时释放相关的软硬件资源，也可以改善用户体验。通过 setOnErrorListener(android.media.MediaPlayer.OnErrorListener)可以设置该监听器。如果 MediaPlayer 进入了 Error 状态，可以通过调用 reset()来恢复，使得 MediaPlayer 重新返回到 Idle 状态。

6.2 音 频 播 放

关于 Android 中音频和视频的播放我们最先想到的就是 MediaPlayer 类，Android 通过 MediaPlayer 类支持播放音频和视频。MediaPlayer 类处于 Android media 包的核心位置，是播放媒体文件最为广泛使用的类。MediaPlayer 已设计用来播放大容量的音频文件以及同样可支持播放操作（停止、开始、暂停等）和查找操作的流媒体，其还可支持与媒体操作相关的监听器。通过以下 3 种方式可完成 MediaPlayer 中的音频播放：

- 从源文件播放。
- 从文件系统播放。
- 从流媒体播放。

6.2.1 从源文件播放音频

使用 MediaPlayer 播放音频或视频，是播放音频文件最普通的方法。在此情况下，音频文件应存在于该项目的 raw 或 assets 文件夹中，如图 6-2 所示。

图 6-2 res 文件夹下的音频文件

要想实现该功能需要以下步骤：

（1）在项目的 res、raw 文件夹下面放一个 Android 所支持的音频文件，例如一个 MP3 文件。

（2）创建一个 MediaPlayer 实例，可以使用 MediaPlayer 的静态方法 create()来完成。

（3）调用 start()方法开始播放音频文件，调用 pause()方法暂停播放，调用 stop()方法停止播放。如果希望重复播放，就必须在调用 start()方法之前，调用 reset()和 prepare()方法。

程序代码如下：

1. `MediaPlayer mPlayer = MediaPlayer.create(this, R.raw.paradoxmp3);`
2. `//开始播放`
3. `mPlayer.start();`

以此类推，可以借鉴 start()，调用其他方法。如要访问一个源资源，仅需使用无扩展名的小写文件名称：

1. `context appContext = getApplicationContext();`
2. `MediaPlayer mPlayer = MediaPlayer.create(appContext,R.raw.paradoxp3);`
3. `mPlayer.start();`

6.2.2 从文件系统播放音频

访问音频文件的第二种方法是从文件系统播放，即 SD 卡。大多数音频资源均存在于 SD 卡中。在研究如何通过 SD 卡访问音频文件之前，让我们看一下如何在 SD 卡中加载文件：

选择 Window → Show View → Other 命令，可打开 Eclipse IDE 中的 FileExplorer 视图。打开 Show View 对话框，如图 6-3 所示，选择 Android → File Explorer 选项。

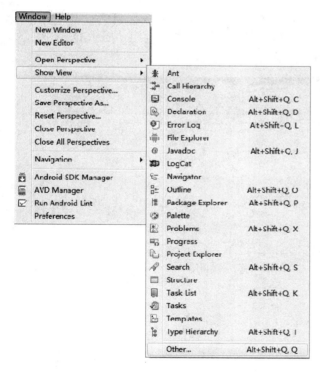
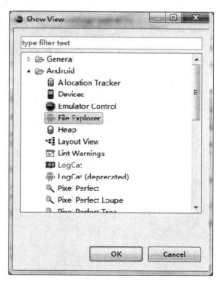

图 6-3　文件加载步骤

一旦选择 File Explorer（文件管理器），即将会打开 File Explorer 视图，如图 6-4 所示。

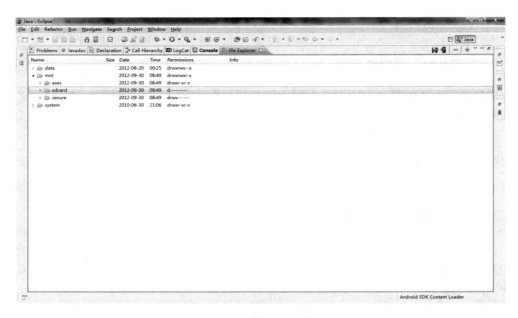

图 6-4 File Explorer 视图

现在,可将文件放入 SD 卡中,在 File Explorer 中选择 sdcard 文件夹,并使用位于右上角的右箭头来选择。此操作可开启对话框,可在其中选择需要上传至 SD 卡中的文件。将文件放入 SD 卡中后,如图 6-5 中显示了可用的内容。

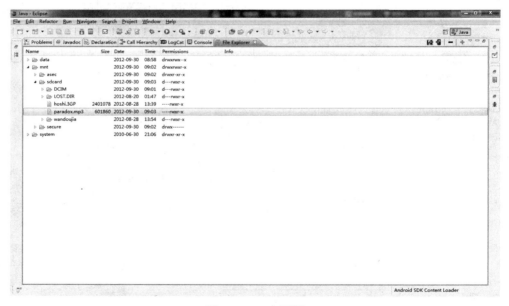

图 6-5 SD 卡视图

学会了如何加载文件后,再开始编写代码。从文件系统中播放音频需要如下步骤:
(1)创建一个新的 MediaPlayer 实例。
(2)调用 setDataSource()方法来设置想要播放的文件路径。

（3）在播放器开始播放音频之前，必须准备好 MediaPlayer 对象。
（4）首先调用 prepare()，然后调用 start() 进行播放。

代码如下：

```
1.   //创建一个新的 MediaPlayer 实例
2.   Mediaplayer mPlayer = new MediaPlayer();
3.   //设置播放路径
4.   String path = "/sdcard/paradox.mp3"
5.   //为 MediaPlayer 设置数据源
6.   mPlayer.setDataSource(path);
7.   mPlayer.prepare();
8.   mPlayer.start();
```

prepare()方法为阻塞方法，并可阻塞直至媒体播放器准备播放歌曲。非阻塞方法 prepareAsync()也可实现。如果媒体播放器用来从流媒体中播放歌曲，并且在播放歌曲之前需要缓冲数据，则应使用非阻塞方法 prepareAsync()。现在使用以下内容来播放控制方法，如 Start()、stop()等。在可设置用于部分其他歌曲文件之前，媒体播放器对象须进行重置。媒体播放器在其使用后须予以释放。此操作使用 release()方法来完成。release()方法可释放与 MediaPlayer 对象相关联的资源。当使用 MediaPlayer 来完成操作时，这被认为是调用此方法的最佳实践。也可通过以下方式来创建媒体播放器：

```
1.   String pathToFile = "/sdcard/paradox.mp3";
2.   MediaPlayer filePlayer = MediaPlayer.create( appContext, Uri.parse
     (pathToFile));
```

此处可通过解析给定的已编译 URI 字符串来使用 URI 类创建 Uri。

6.2.3 从流媒体播放音频

大家都应该有这样的经历：一打开某个网页，有时就会有音乐响起。例如说 QQ 空间的 QQ 音乐，进入某某好友或自己的 QQ 空间（前提是设置了背景音乐的），网页就会自己播放音乐。下面就来看看如何通过网络来播放音频文件。

使用与用于访问 SD 卡中音频文件的相同代码，可访问网站中的音频文件。唯一的变化就是文件路径。此处的路径为网站 URL，其指向音频资源文件。此处最重要的部分就是使用互联网提取数据，因此必须获取访问互联网的许可。在 AndroidManifest.xml 文件中设置互联网许可：

```
<uses-permission android:name="android.permission.INTERNET">
    </uses-permission>
```

除了 URL 路径外，其他部分代码保持相同。URL 的路径应该设置为：

```
String urlPath = "http://www.XXX.com/…/paradox.mp3";//或是其他网络地址
```

也可以通过下面的方式创建媒体播放器：

```
1.  String urlPath = "http:/www.XXX.com/…/paradox.mp3";
2.  MediaPlayer filePlayer = MediaPlayer.create( appContext, Uri.parse
    (urlPath));
```

此处可通过解析给定的已编译 URI 字符串来使用 URI 类创建 Uri。

6.3 视 频 播 放

通过上面的学习，相信大家已经对 MediaPlayer 类以及媒体播放有了了解，下面进一步学习媒体播放中的视频播放。

Android 提供了专业化的视图控制 android.widget.VideoView，其可压缩创建并初始化 MediaPlayer。VideoView 类可从各种源（如资源或内容提供商）加载图片，并且可负责从该视频计算其尺寸，以便其可在任何布局管理器中使用。同样，该类还可提供各种显示选项，如缩放比例和着色。可用来显示 SDCard FileSystem 中存在的视频文件或联机存在的文件。

与音频播放一样，视频播放也有 3 种方式。

6.3.1 从源文件播放视频

如果想要从源文件里播放视频，需要使用 MediaPlayer 静态方法 MediaPlayer create (Context context, int resid)来创建 MediaPlayer 对象。该媒体播放器要求一个 SurfaceHolder 对象来显示视频内容，该对象可使用 setDisplay()方法来予以分配。程序代码如下：

```
1.  //创建一个 MediaPlayer 实例
2.  MediaPlayer mPlayer = MediaPlayer.create(this, R.raw.samplemp4);
3.  //设置用 SurfaceHolder 来显示多媒体
4.  mPlayer.setDisplay(holder);
5.  mPlayer.setAudioStreamType(AudioManager.STREAM_MUSIC);
6.  //开始播放
7.  mPlayer.start();
```

6.3.2 从文件系统播放视频

要从文件系统播放视频，需要使用 setDataSource (String path)方法来将文件路径设置为 MediaPlayer。然后，使用 prepare()方法来准备 MediaPlayer。prepare()方法可同步准备播放器进行播放。设置 datasource 和显示工具后，需要调用 prepare()或 prepareAsync()。因为特定的方法（如 setDataSource()、prepare()）可能会抛出异常，所以要使用 try-catch Exception 语句。

```
1.  String pathToFile = "/sdcard/paradox.mp4";
2.  // 创建一个 MediaPlayer 实例
3.  MediaPlayer mMPlayer = new MediaPlayer();
4.  //设置路径
```

```
5.  mPlayer.setDataSource(pathToFile);
6.  //设置用 SurfaceHolder 来显示多媒体
7.  mPlayer.setDisplay(holder);
8.  //播放准备
9.  mPlayer.prepare();
10. mPlayer.setAudioStreamType(AudioManager.STREAM_MUSIC);
11. //开始播放
12. mPlayer.start();
```

6.3.3 从流媒体播放视频

在网上看电视剧、电影等视频时，你有没有想过，视频是怎样在网页上播放的？其实，这不是很复杂的事。前面学了如何在文件系统中播放视频，这里只需要将文件的路径更改一下，设置至可访问视频的网站，其他的保持不变就可以了。例如：

```
String pathToFile = "http://www.XXX.com/.../samplemp4.mp4";
```

很容易看出，这与在网页上播放音频文件的路径设置方式是一样的。

6.4 为图书管理系统配上音乐

本章讲解的是多媒体的知识，本节将在第 5 章 TCP 图书管理系统的基础上，增加一小段播放音乐的代码，为图书管理系统配上音乐。首先运行 Chapter06，效果如图 6-6 和图 6-7 所示。

图 6-6　Chapter06 运行效果

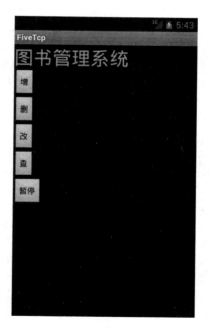
图 6-7　单击"播放"按钮后

Chapter06 在 res 下的 raw 文件夹中添加了 music01.mp3 文件，如图 6-8 所示。

Chapter06 在 Chapter05_tcp 的 ui.cqupt 包中的 MainActivity 添加了少量代码，添加了一个播放按钮。Chapter06 的 MainActivity 代码如下所示：

图 6-8　添加的 mp3 文件

```
1.  package ui.cqupt;
2.
3.  import control.cqupt.Controller;
4.  import net.cqupt.Client;
5.  import android.app.Activity;
6.  import android.content.Intent;
7.  import android.media.MediaPlayer;
8.  import android.os.Bundle;
9.  import android.view.View;
10. import android.view.View.OnClickListener;
11. import android.widget.Button;
12.
13. public class MainActivity extends Activity {
14. private Button play;
15. private MediaPlayer mediaPlayer;
16.     public void onCreate(Bundle savedInstanceState) {
17.         super.onCreate(savedInstanceState);
18.         setContentView(R.layout.main);
19.         Button insert = (Button) findViewById(R.id.m_insert);
20.         Button delete = (Button) findViewById(R.id.m_delete);
21.         Button set = (Button) findViewById(R.id.m_set);
22.         Button select = (Button) findViewById(R.id.m_select);
23.         play=(Button) findViewById(R.id.m_play);
24.         mediaPlayer=MediaPlayer.create(this, R.raw.music01);
25.         ButtonListener buttonListener = new ButtonListener();
26.         insert.setOnClickListener(buttonListener);
27.         delete.setOnClickListener(buttonListener);
28.         set.setOnClickListener(buttonListener);
29.         select.setOnClickListener(buttonListener);
30.         play.setOnClickListener(buttonListener);
31.     }
32.
33.     public void onDestroy() {
34.         mediaPlayer.release();
35.         mediaPlayer = null;
36.         Controller control = new Controller();
37.         control.exit();
```

```
38.         Client.close();
39.         super.onDestroy();
40.     }
41.
42.     class ButtonListener implements OnClickListener {
43.
44.         public void onClick(View v) {
45.             int id = v.getId();
46.             Intent intent = new Intent();
47.             switch (id) {
48.             case R.id.m_insert:
49.                 intent.setClass(MainActivity.this, InsertActivity.class);
50.                 MainActivity.this.startActivity(intent);
51.                 break;
52.             case R.id.m_delete:
53.                 intent.setClass(MainActivity.this, DeleteActivity.class);
54.                 MainActivity.this.startActivity(intent);
55.                 break;
56.             case R.id.m_set:
57.                 intent.setClass(MainActivity.this, SetActivity.class);
58.                 MainActivity.this.startActivity(intent);
59.                 break;
60.             case R.id.m_select:
61.                 intent.setClass(MainActivity.this, SelectActivity.class);
62.                 MainActivity.this.startActivity(intent);
63.                 break;
64.             case R.id.m_play:
65.                 if("播放".equals(play.getText().toString()))
66.                 {
67.                     mediaPlayer.start();
68.                     play.setText("暂停");
69.
70.                 }else if("暂停".equals(play.getText().toString()))
71.                 {
72.                     mediaPlayer.pause();
73.                     play.setText("播放");
74.                 }
75.                 break;
76.             }
77.         }
```

```
78.
79.     }
80. }
```

首先在第 14 行声明了一个 Button 组件,并在第 23 行通过 findViewById 方法实例化 Button 组件,在第 30 行为这个 Button 组件添加监听器。此 Button 组件在 main.xml 文件中有声明,代码如下:

```
1.  <Button
2.      android:id="@+id/m_play"
3.      android:layout_width="wrap_content"
4.      android:layout_height="wrap_content"
5.      android:text="播放"
6.  />
```

在第 15 行声明了一个 MediaPlayer 对象,在第 24 行通过 MediaPlayer 类的 create 方法指定保存在 res\raw 文件夹中的 mp3 资源,并实例化 MediaPlayer 对象。当用户单击"播放"按钮时,则执行第 67 行调用 MediaPlayer 的 start 方法开始播放音乐,在第 68 行把"播放"按钮上的文字改为暂停。如果继续单击按钮,则执行第 72 行 MediaPlayer 的 pause 方法暂停音乐,然后在第 73 行将按钮上的字体设置成"播放"。在退出应用程序的时候最好执行第 34 行 MediaPlayer 的 release 方法释放音频文件,并且在第 35 行将 MediaPlayer 对象设置为 null。

第 7 章 图书管理系统程序进阶

Android 作为一个优秀的操作系统，拥有非常丰富的功能，除了之前讲过的功能外，我们将在本章中讲解 Android 的其他一些重要功能。

7.1 Service（服务）

第 1 章介绍了 Android 的四大组件，对 Service 的基本功能有所了解，本节将重点介绍 Service 是什么、Service 的启动及生命周期。让读者对 Service 有更清晰的理解。

7.1.1 了解 Service

1. Service 是什么

Service 是 Android 的四大组件之一，但与 Activity 不同的是，Service 是不可见的，它没有 Activity 那样丰富的界面，用户一般感觉不到它的存在，因为它是后台运行的。Service 用于处理那些比较耗时或需要长时间运行的操作，例如，播放背景音乐。我们还可以使用 Service 实现更新 ContentProvider、发送 Intent 以及启动系统通知的功能。

2. Service 不是什么

通过上面的描述我们知道了什么是 Service，但我们又不禁猜想 Service 是否就是一个线程呢？下面就告诉大家 Service 不是什么。

- Service 不是一个进程。除非特别对 Service 对象声明要以单独进程运行，Service 对象是不会以单独的进程运行的。
- Service 不是一个线程，Service 不会脱离主线程单独运行。

这些知识在 Android 的官方文档都有详细的讲解，有兴趣的读者可以去查看。

7.1.2 Service 的启动与生命周期

1. Service 的启动

Service 的启动可分为两种形式：

- 显示调用。当一个程序组件调用了 startService()方法时，一个 Service 就会被显示启动，在后台运行时该组件被终止，这个 Service 仍然继续运行。这种调用方式通常用于那种不需要向调用者返回结果的情况，比如说在网络下载、上传文件。操作完成服务应被停止。
- 绑定调用。Service 绑定是通过调用 bindService()方法实现的。采用绑定的方式启动 Service 允许组件通过 client-server 接口与 Service 交互，甚至完成进程通信（将在 7.2 节介绍）。多个组件可以同时绑定一个组件，只有在所有绑定解除时这个 Service 才会停止。

2. Service 的生命周期

Service 因为拥有两种不同的启动形式，所以拥有两种不同的生命周期，采用的启动形式不同，其生命周期必然不同。下面就来看看两种启动形式下的生命周期。

先看一张 Android 官方 API 中展示的 Service 生命周期的图解（见图 7-1）。

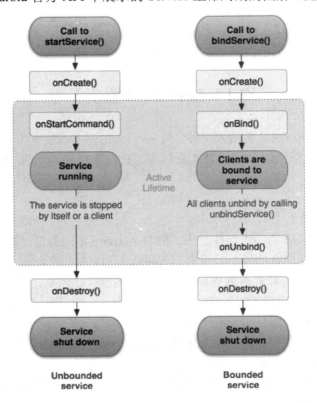

图 7-1 Service 的两种生命周期

1）显式调用

Android 提供了两个显式调用方法分别控制 Service 的启动与停止。

- startService(Intent service)：启动服务。
- stopService(Intent service)：停止服务。

显示调用生命周期分为创建、开始和销毁 3 个阶段。对应 Service 的 3 个函数：onCreate()→onStartCommand ()→onDestroy()。

下面通过一段简单的代码来观察 Service 的这种生命周期。

（1）将 Service 定义为以下格式：

```
1.  public class Service01 extends Service{
2.      public IBinder onBind(Intent intent) {
```

```
3.            Log.d("Service", "onBind");
4.            return null;
5.        }
6.        public void onCreate() {
7.            Log.d("Service", "onCreate");
8.            super.onCreate();
9.        }
10.       public int onStartCommand(Intent intent, int flags, int startId) {
11.           Log.d("Service", "onStartCommand");
12.           return super.onStartCommand(intent, flags, startId);
13.       }
14.       public void onDestroy() {
15.           Log.d("Service", "onDestroy");
16.           super.onDestroy();
17.       }
18. }
```

（2）建立一个 Activity，上面放置两个按钮：startServiceBtn 和 stopServiceBtn，并设置按钮的响应函数。

```
1.  public class MainActivity extends Activity {
2.      private Button startServiceBtn;
3.      private Button stopServiceBtn;
4.      public void onCreate(Bundle savedInstanceState) {
5.          super.onCreate(savedInstanceState);
6.          setContentView(R.layout.main);
7.          startServiceBtn = (Button) findViewById(R.id.startService);
8.          startServiceBtn.setOnClickListener(new
                OnstartServiceListener());
9.          stopServiceBtn = (Button) findViewById(R.id.stopService);
10.         stopServiceBtn.setOnClickListener(new
                OnstopServiceListener());
11.     }
12.     private class OnstartServiceListener implements OnClickListener
13.     {
14.         public void onClick(View arg0) {
15.             Intent intent = new Intent();
16.             intent.setClass(MainActivity.this, Service01.class);
17.             startService(intent);
18.         }
19.     }
20.     private class OnstopServiceListener implements OnClickListener {
21.         public void onClick(View arg0) {
22.             Intent intent = new Intent();
```

```
23.             intent.setClass(MainActivity.this, Service01.class);
24.             stopService(intent);
25.         }
26.     }
27. }
```

程序运行界面如图 7-2 所示。

图 7-2 显示启动 Service 生命周期测试

单击 startService 按钮时，DDMS 中 Log 打印信息如图 7-3 所示。

```
10-02 0... D  367  Service  onCreate
10-02 0... D  367  Service  onStartCommand
```

图 7-3 单击 startService 按钮

重复单击 startService 按钮时，DDMS 中 Log 打印信息如图 7-4 所示。

```
10-02 0... D  367  Service  onCreate
10-02 0... D  367  Service  onStartCommand
10-02 0... D  367  Service  onStartCommand
10-02 0... D  367  Service  onStartCommand
10-02 0... D  367  Service  onStartCommand
10-02 0... D  367  Service  onStartCommand
10-02 0... D  367  Service  onStartCommand
```

图 7-4 重复单击 startService 按钮

单击 stopService 按钮，DDMS 中 Log 打印信息如图 7-5 所示。

```
10-02 0... D  367  Service  onDestroy
```

图 7-5 单击 stopService

综上可知：

在 Service 的整个生命周期中，onCreate()和 onDestroy()函数只会在 Service 创建和销毁时被调用一次，无论调用多少次 startService 方法，Service 只会被创建一次，被重复调用的是 onStartCommand()方法。

本部分测试所用代码见本书配套资源中的工程 Chapter7.1.1。

2）绑定启动

Android 中提供了一个方法将 Service 和 Activity 绑定起来。

bindService(Intent service,ServiceConnectoin conn,int flags)：第一个参数表示与服务类相关联的 Intent 对象，第二个参数是一个 ServiceConnection 类型的变量，负责连接 Intent 对象指定的服务。通过 ServiceConnection 对象可以获得连接成功或者失败的状态，并可以获得连接后的服务对象。第三个参数是一个标志位，设为 Context.BIND_AUTO_CREATE 即可。

这里简单地体验一下通过绑定启动的 Service 的生命周期。

（1）定义 Service，代码如下：

```
1.   public class MainActivity extends Activity {
2.       private Button bindServiceBtn;
3.       private Button stopServiceBtn;
4.       private Service01 mService;
5.       private ServiceConnection serviceConnection = new
         ServiceConnection(){
6.           public void onServiceDisconnected(ComponentName name){
7.               Log.d("Service", "Service Failed");
8.           }
9.           public void onServiceConnected(ComponentName name, IBinder
             service){
10.              MyBinder binder = (Service01.MyBinder) service;
11.            mService = binder.getService();
12.              Log.d("Service", "Service Connected");
13.          }
14.      };
15.      public void onCreate(Bundle savedInstanceState) {
16.          super.onCreate(savedInstanceState);
17.          setContentView(R.layout.main);
18.          bindServiceBtn = (Button) findViewById(R.id.bindService);
19.          bindServiceBtn.setOnClickListener(new
             OnstartServiceListener());
20.          stopServiceBtn = (Button) findViewById(R.id.stopService);
21.          stopServiceBtn.setOnClickListener(new
             OnstopServiceListener());
22.      }
23.      private class OnstartServiceListener implements OnClickListener
24.      {
```

```
25.        public void onClick(View arg0) {
26.            Intent intent = new Intent();
27.            intent.setClass(MainActivity.this, Service01.class);
28.            bindService(intent, serviceConnection, Context.BIND_
               AUTO_CREATE);
29.        }
30.    }
31.    private class OnstopServiceListener implements OnClickListener {
32.        public void onClick(View arg0) {
33.            Intent intent = new Intent();
34.            intent.setClass(MainActivity.this, Service01.class);
35.            unbindService(serviceConnection);
36.        }
37.    }
38. }
```

（2）创建一个 Activity，放置两个按钮：bindServiceBtn 和 UnbindServiceBtn，并分别添加事件响应，代码如下：

```
1.  public class Service01 extends Service {
2.      private MyBinder myBinder = new MyBinder();
3.      public void onCreate() {
4.          Log.d("Service", "onCreate");
5.          super.onCreate();
6.      }
7.      public IBinder onBind(Intent intent) {
8.          Log.d("Service", "onBind");
9.          return myBinder;
10.     }
11.     public void onDestroy() {
12.         Log.d("Service", "onDestroy");
13.         super.onDestroy();
14.     }
15.     public void onRebind(Intent intent) {
16.         Log.d("Service", "onRebind");
17.         super.onRebind(intent);
18.     }
19.     public boolean onUnbind(Intent intent) {
20.         Log.d("Service", "onUnbind");
21.         return super.onUnbind(intent);
22.     }
23.     public class MyBinder extends Binder {
24.         Service01 getService() {
25.             return Service01.this;
```

```
26.         }
27.     }
28. }
```

程序运行界面如图 7-6 所示。

图 7-6 绑定的 Service 生命周期测试

当单击 bindService 按钮时，DDMS 中 Log 打印信息如图 7-7 所示。

```
10...   D   481   S...   onCreate
10...   D   481   S...   onBind
10...   D   481   S...   Service Connected
```

图 7-7 单击 bindService

然后再单击 unbindService 按钮，DDMS 中 Log 打印信息如图 7-8 所示。

```
10...   D   481   S...   onUnbind
10...   D   481   S...   onDestroy
```

图 7-8 单击 unbindService

这种启动方式所对应的就是图 7-1 的第二种生命周期。

在这段代码中，ServiceConnection 这个对象是用于建立连接的。当 Service 与 Activity 绑定成功时，链接建立，此时会运行 onServiceConnected()中的代码；当连接意外断开时，会运行 onServiceDisconnected()中的代码。

当 Activity 开始运行时，系统将会调用 onBind()函数使 Service 开始运行；当 Activity 停止时，系统会调用 onUnbind()函数将 Activity 与 Service 之间的连接断开。

绑定操作通常除了用来控制 Service 的启动与停止，还常常被用来进行数据的传递。

其他进程在绑定 Service 的同时，还可以通过 ServiceConnection 和 bindService()方法的结合使用，在绑定的同时获取这个 Service 的对象或者 Service 内的某些对象、数据。

本部分测试所用代码见本书配套资源中的工程 Chapter7.1.2。

7.2 系统服务

Android 系统中有很多内置软件，如电话、短信、重力感应等。除了系统调用和用户直接使用外，也可以通过系统提供的接口来调用这些服务，使自己的程序更加精彩。本节主要介绍系统服务的用法。

7.2.1 什么是系统服务

系统服务是一种 Service，只不过这些服务是由系统自主运行，并为系统的功能提供支持。普通的 Service 一般只是为程序员编写的一些软件提供服务，而系统服务则是提供系统中的某些功能，如电话、短信和感应器等。

7.2.2 获得系统服务

一个系统服务其实就是一个对象，通过 Activity 类的 getSystemService()方法可以获得指定的系统服务对象。getSystemService()方法只有一个 String 类型的参数，表示系统服务的 ID，这个 ID 在整个 Android 系统中是唯一的。例如，audio 表示音频服务，window 表示窗口服务，notification 表示通知服务。为了便于记忆和管理，Android SDK 在 android.content.Context 类中定义了这些 ID，如：

- public static final String AUDIO_SERVICE="audio"——音频服务的 ID。
- public static final String WINDOW_SERVICE="window"——窗口服务的 ID。

获取了服务对象后，就可以使用系统提供的方法对服务进行操作或者获取数据了。例如下面这段代码：

```
1.  AudioManager audio=(AudioManager)getSystemService(Context.AUDIO_
    SERVICE);
2.  audio.setRingerMode(AudioManager.RINGER_MODE_SILENT);
```

第 1 行代码获得了系统的音频服务，使用第 2 行代码是使手机静音。

表 7-1 列举了常用的系统服务及其 ID。

表 7-1　常用的系统服务

服务 ID	返回的对象	服务名称
WINDOW_SERVICE	WindowManager	管理打开的窗口程序
POWER_SERVICE	PowerManger	电源的服务
ALARM_SERVICE	AlarmManager	闹钟的服务
KEYGUARD_SERVICE	KeyguardManager	键盘锁的服务
LOCATION_SERVICE	LocationManager	位置服务，如 GPS

续表

服务 ID	返回的对象	服务名称
VIBRATOR_SERVICE	Vibrator	手机震动的服务
CONNECTIVITY_SERVICE	Connectivity	网络连接的服务
WIFI_SERVICE	WifiManager	WiFi 服务
DOWNLOAD_SERVICE	DownloadService	网络下载服务
INPUT_METHOD_SERVICE	InputMethodManager	输入法服务

系统服务的知识相当多，如果读者想了解更多有关的系统服务，可以查阅 Android 官方 API。

7.2.3 重力感应

在 Android 系统提供了众多的系统服务，例如音频服务、WiFi 服务、重力感应等，在众多的系统服务中，重力感应这一功能一直是智能手机的一大卖点，这一功能衍生出了许多操作性强、操作快捷和有特色的应用。这里以重力感应作为系统服务的代表进行讲解。

重力感应功能在 Android 中并不是一个独立的服务进程，而是作为传感器中的一种感应功能。所以要获得重力感应的对象，首先需要获得传感器的系统服务，代码如下：

```
SensorManager sensor= (SensorManager) getSystemService(SENSOR_SERVICE);
```

传感器中包含了多种感应功能，表 7-2 列举了感应器的部分功能。

表 7-2 感应器的部分功能

感应功能 ID	数据类型	注
Int	TYPE_ACCELEROMETER	加速度
Int	TYPE_ALL	所有类型，在 NexusOne 中默认为加速度
Int	TYPE_GYROSCOPE	回转仪
Int	TYPE_LIGHT	光线感应
Int	TYPE_ORIENTATION	指北针和角度

这里只是列举了部分功能，如需获取更详细的内容，请查阅 Android 官方 API。获取重力感应可以直接使用 TYPE_ALL 参数：

```
Sensor sensor = sensorManager.getDefaultSensor(Sensor.TYPE_ALL);
```

获取重力感应对象后，可以通过 values[]方法获得手机 XYZ 三轴的偏转角度。三轴的值为 float 类型，取值范围为 –10～10。

表 7-3 手机放置状态与 XYZ 三轴的值的关系

手机放置状态	X 轴	Y 轴	Z 轴
手机屏幕向上（Z 轴朝天）水平放置	0	0	10
手机屏幕向下（Z 轴朝地）水平放置	0	0	–10

续表

手机放置状态	X 轴	Y 轴	Z 轴
手机横向放置，屏幕与地面垂直（X 轴朝天）	10	0	0
手机竖直（Y 轴朝天）向上	0	10	0

可以添加重力感应的监听，当手机的放置状态发生改变时，就能够获取新的手机放置信息：

```
1.  float x,y,z;
2.  SensorEventListener sel=new SensorEventListener(){
3.       public void onSensorChanged(SensorEvent se) {
4.           x=se.values[SensorManager.DATA_X];
5.           y=se.values[SensorManager.DATA_Y];
6.           z=se.values[SensorManager.DATA_Z];
7.       }
8.       public void onAccuracyChanged(Sensor arg0, int arg1) {
9.       }
10.  };
11. }
```

代码中 x、y、z 分别代表 X、Y、Z 的信息，获取到手机放置信息后，就可以通过 XYZ 轴的值来进行我们需要的操作了。

7.3 广　　播

7.3.1 什么是广播

1. 日常生活中的广播（Broadcast）

大家对日常生活中的广播应该有一定的了解，例如收音机广播。每个广播台播放的内容都不同，接收广播时广播（发送者）不会在意我们接收到广播后如何处理，就如同在开车时广播告诉我们前方道路堵塞，但是它不会管我们该怎么处理——是等待道路疏通还是换道。那么在 Android 中广播机制是怎样的呢？

2. Android 中的广播

Android 中有各种各样的广播，例如电池的使用状态、电话的接收和短信的接收都会产生一个广播。各种广播在 Android 系统中运行，当系统/应用程序运行时便会向 Android 注册各种广播，Android 接收到广播会便会判断哪种广播需要哪种事件，然后向需要不同事件响应的应用程序注册事件，不同的广播可能处理不同的事件，也可能处理相同的广播事件，这时就需要 Android 系统为我们做筛选。

在 Android 中，广播的生命周期也很简单，下面来看一下 API 给出的广播生命周期的解释，如图 7-9 所示。

图 7-9 的大概意思是：如果一个广播处理完 onReceive，那么系统将认定此对象将不

再是一个活动的对象，也就会结束它。

> **Receiver Lifecycle**
> A BroadcastReceiver object is only valid for the duration of the call to onReceive(Context, Intent). Once your code returns from this function, the system considers the object to be finished and no longer active.

图 7-9 广播的 API 解释

7.3.2 广播的接收与响应

如果想在 Android 中接收广播信息，那么首先需要有一个广播接收器，而这个广播接收器需要我们自己来实现。只要编写一个类，然后继承 BroadcastReceiver，就可以得到一个广播接收器了。然后要重写 BroadcastReceiver 中的 onReceiver 方法，该方法的作用就是定义当某条广播到来时，告诉我们应该怎么处理。还需要在 AndroidManifest.xml 文件中使用<receiver>标签来指定继承接收系统广播的类可以接收哪一个 Broadcast Action，代码如下：

```
1.   <receiver android:name=" StartReceiver ">
2.       <intent-filter>
3.           <action android:name="android.intent.action.BOOT_COMPLETED"/>
4.       </ intent-filter >
5.   </receiver >
```

其中的<action>标签就是用来指定接收的广播种类的，StartReceiver 在这里为指定开机启动的广播（如需更多参数，可以查阅 Android API）。

一个关于广播的小程序

在这个小程序中，我们想要实现开机启动一个 Activity。首先需要接收到开机这个系统广播，然后再在上面讲到的 onReceiver 方法中打开想要启动的 Activity。代码如下：

```
1.   public class StartReceiver extends BroadcastReceiver{
2.       public void onReceive(Context context, Intent intent){
3.           Intent mainIntent = new Intent(context, Main.class);
4.           mainIntent.setFlags(Intent.FLAG_ACTIVITY_NEW_TASK);
5.           context.startActivity(mainIntent);
6.       }
7.   }
8.   public class Main extends Activity{
9.       public void onCreate(Bundle savedInstanceState){
10.          super.onCreate(savedInstanceState);
11.          setContentView(R.layout.main);
12.      }
13.  }
```

详细代码见本书配套资源中的工程 Chapter7.3.2。

最后在 AndroidManifest.xml 文件中配置好 Broadcast 的<receiver>标签。这里使用的就是本节开始的那段配置。配置完成后，一个简单的开机启动程序就完成了。运行程序后，当系统启动时，Activity 就会自动启动。如图 7-10 所示。

7.3.3 广播的发送

在 Android 中并不是只有系统才能够发送广播，当程序员需要通知其他的应用程序或向其他应用程序发送数据时，也可以考虑使用 **sendBroadcast()** 方法发送广播。当然在发送了广播后，使用 7.3.2 节的方法即可接收到自己所设定的广播。

下面是发送广播的小实例，该程序定义了一个名为 SELF_BROADCAST 的广播。代码如下：

图 7-10　广播实现的开机启动

```
1.  public class Main extends Activity {
2.      Intent broadcastIntent =new Intent("SELF_BROADCAST");
3.      class ButtonListener implements OnClickListener {
4.        public void onClick(View view){
5.          switch (view.getId()){
6.            case R.id.sendBroadcast:
7.                sendBroadcast(broadcastIntent);
8.                break;
9.          }
10.       }
11.     };
12.     public void onCreate(Bundle savedInstanceState){
13.         super.onCreate(savedInstanceState);
14.         setContentView(R.layout.main);
15.         Button sendBroadcast = (Button) findViewById
            (R.id.sendBroadcast);
16.         sendBroadcast.setOnClickListener(this);
17.     }
18. }
```

详细代码见本书配套资源中的工程 Chapter7.3.3。

这里广播被设为了 SELF_BROADCAST，只需将接收广播的进程中<receiver>标签下的<action>标签的 android:name 属性设定为 SELF_BROADCAST 即可。当然，也可以设定为其他没有被系统使用的值。

图 7-11 为发送广播的界面，单击"发送"按钮后程序运行，进行接收广播后的响应。如图 7-12 所示。

图 7-11 发送广播的界面

图 7-12 接收到广播后

7.4 Service 实现新书上架通知

7.4.1 客户端

本节运用所学的 Service 功能实现新书上架通知的功能，主要运用了 Service 的知识，本节工程 Chapter07_Client 是在 Chapter06_tcp 的基础上增加了新功能。下面来看一下 Chapter07_Client 的包结果，如图 7-13 所示。

由图 7-13 可以看到，与 Chapter06_tcp 相比较，Chapter07_Client 工程在 net.cqupt 包中多了 UpDateService 类，这就是 Service 类。下面就开始对 UpDateService 进行分析，UpDateService 的代码如下所示：

```
1.  package net.cqupt;
2.
3.  import java.io.IOException;
4.  import java.io.InputStream;
5.  import java.net.Socket;
6.  import control.cqupt.Controller;
7.  import android.app.Service;
8.  import android.content.Intent;
9.  import android.os.Binder;
10. import android.os.Handler;
11. import android.os.IBinder;
12.
13. public class UpDateService extends Service {
```

图 7-13 包结构

```
14.
15.     private MyBinder myBinder = new MyBinder();
16.
17.     public IBinder onBind(Intent intent) {
18.         return myBinder;
19.     }
20.
21.     public boolean onUnbind(Intent intent) {
22.         return super.onUnbind(intent);
23.     }
24.
25.     public void update(Handler handler) {
26.         new ServiceThread(handler).start();
27.     }
28.
29.     public class ServiceThread extends Thread {
30.         private Handler handler;
31.
32.         public ServiceThread(Handler handler) {
33.             this.handler = handler;
34.         }
35.
36.         public void run() {
37.             try {
38.                 Socket socket = new Socket("172.22.146.112", 8003);
39.                 while (true) {
40.                     InputStream in = socket.getInputStream();
41.                     byte[] buffer = new byte[1024];
42.                     int index = in.read(buffer);
43.                     String message = new String(buffer, 0, index, "GBK");
44.                     Controller controller = new Controller(handler);
45.                     controller.doUpdate(message);
46.                 }
47.             } catch (IOException e) {
48.                 e.printStackTrace();
49.             }
50.         }
51.     }
52.
53.     public class MyBinder extends Binder {
54.         public UpDateService getService() {
55.             return UpDateService.this;
56.         }
```

```
57.     }
58. }
```

Chapter07_Client 使用的是绑定 Activity 的方式启动 Service，在 UpDateService 类中覆写了第 17~23 行的两个方法 onBind 和 onUnbind，在 onBind 里返回了在第 15 行定义的 MyBinder 对象，MyBinder 是在第 53~56 行定义的内部类，定义这个类的原因在本章已经讲过，因为在调用 Service 的时候需要用到类型为 Binder 的对象。当启动 Activity 开启 Service 时，就会调用第 25~27 行的 update 方法开启一个线程 ServerThread。第 29~51 行自定义了一个内部类 ServiceThread，这是一个线程类。在这个内部类的构成函数中为它的成员变量 handler 对象赋值，这个 handler 对象更新 MainActivity 主界面。在 ServerThread 的 run 方法里实现了不断接收服务器的消息。在第 38 行通过 socket 建立连接，在第 40~45 行循环接收服务器的消息，接收到的消息通过第 45 行 Controller 中的 doUpdate 方法更新界面。Controller 中 doUpdate 代码如下所示：

```
1.  public void doUpdate(String str)
2.  {
3.      Parser parser=new Parser();
4.      Book book=parser.parseBook(str);
5.      Message msg = new Message();
6.      Bundle bundle = new Bundle();// 存放数据
7.      bundle.putString("content", book.toString());
8.      msg.setData(bundle);
9.      handler.sendMessage(msg);
10. }
```

doUpdate 的方法和之前所讲的 doResponse 的实现是一样的，通过 Parser 解析数据放入 Bundle 中，通过 handler 的 sendMessage 方法发送到界面的消息的队列中。

下面讲解 MainActivity 中启动和调用 Service，代码如下所示：

```
1.  package ui.cqupt
2.  
3.  import control.cqupt.Controller;
4.  import net.cqupt.Client;
5.  import net.cqupt.UpDateService;
6.  import net.cqupt.UpDateService.MyBinder;
7.  import android.app.Activity;
8.  import android.app.AlertDialog.Builder;
9.  import android.content.ComponentName;
10. import android.content.Context;
11. import android.content.DialogInterface;
12. import android.content.Intent;
13. import android.content.ServiceConnection;
14. import android.os.Bundle;
```

```
15. import android.os.Handler;
16. import android.os.IBinder;
17. import android.os.Message;
18. import android.view.View;
19. import android.view.View.OnClickListener;
20. import android.widget.Button;
21.
22. public class MainActivity extends Activity {
23.     private UpDateService upDateService;
24.
25.     public void onCreate(Bundle savedInstanceState) {
26.         super.onCreate(savedInstanceState);
27.         setContentView(R.layout.main);
28.         Button insert = (Button) findViewById(R.id.m_insert);
29.         Button delete = (Button) findViewById(R.id.m_delete);
30.         Button set = (Button) findViewById(R.id.m_set);
31.         Button select = (Button) findViewById(R.id.m_select);
32.         ButtonListener buttonListener = new ButtonListener();
33.         insert.setOnClickListener(buttonListener);
34.         delete.setOnClickListener(buttonListener);
35.         set.setOnClickListener(buttonListener);
36.         select.setOnClickListener(buttonListener);
37.         Intent bindIntent = new Intent(this, UpDateService.class);
38.         this.getApplicationContext().bindService(bindIntent, serviceConnection,
39.             Context.BIND_AUTO_CREATE);
40.     }
41.
42.     public void onDestroy() {
43.         this.getApplicationContext().unbindService(serviceConnection);
44.         Controller control = new Controller();
45.         control.exit();
46.         Client.close();
47.         super.onDestroy();
48.     }
49.
50.     private Handler handler = new Handler() {
51.         @Override
52.         public void handleMessage(Message msg) {
53.             super.handleMessage(msg);
54.             Bundle b = msg.getData();
55.             String content = b.getString("content");
56.             buildDialog(content);
57.         }
```

```
58.        };
59.
60.    private void buildDialog(String result) {
61.        Builder builder = new Builder(this);
62.        builder.setTitle("新增图书" + "\n" + result);
63.        builder.setNegativeButton("查询", new DialogInterface.
           OnClickListener() {
64.            public void onClick(DialogInterface dialog, int
               whichButton) {
65.                Intent intent = new Intent();
66.                intent.setClass(MainActivity.this, SelectActivity.
                   class);
67.                MainActivity.this.startActivity(intent);
68.            }
69.
70.        });
71.        builder.setPositiveButton("取消", null);
72.        builder.show();
73.    }
74.
75.    private ServiceConnection serviceConnection = new
           ServiceConnection() {
76.
77.        public void onServiceConnected(ComponentName name, IBinder
           service) {
78.            MyBinder myBinder = (UpDateService.MyBinder) service;
79.            upDateService = myBinder.getService();
80.            upDateService.update(handler);
81.        }
82.
83.        public void onServiceDisconnected(ComponentName name) {
84.            upDateService = null;
85.        }
86.
87.    };
88.
89.    class ButtonListener implements OnClickListener {
90.
91.        public void onClick(View v) {
92.            int id = v.getId();
93.            Intent intent = new Intent();
94.            switch (id) {
95.            case R.id.m_insert:
96.                intent.setClass(MainActivity.this, InsertActivity.
```

```
97.              MainActivity.this.startActivity(intent);
98.              break;
99.          case R.id.m_delete:
100.             intent.setClass(MainActivity.this, DeleteActivity.
                 class);
101.             MainActivity.this.startActivity(intent);
102.             break;
103.         case R.id.m_set:
104.              intent.setClass(MainActivity.this, SetActivity.
                  class);
105.              MainActivity.this.startActivity(intent);
106.              break;
107.         case R.id.m_select:
108.              intent.setClass(MainActivity.this, SelectActivity.
                  class);
109.              MainActivity.this.startActivity(intent);
110.              break;
111.         }
112.     }
113.
114.    }
115. }
```

MainClass 在 onCreate 方法中第 37～39 行，通过 Intent 启动 Service 并且通过 bindService 方法将 Service 绑定到 MainActivity 上。bindService 中有一个参数是 serviceConnection，这是一个 ServiceConnection 对象，是在第 75～87 行定义的，如果 MainActivity 和 UpDateService 连接成功，就会执行第 77～81 行的 onServiceConnected 方法，在 onServiceConnected 方法中对 UpDateService 进行了操作，在第 78、79 行得到操作 Service 的对象，在第 80 行调用 UpDateService 中的 update 方法并将 handler 对象作为参数传递。如果连接失败了，则执行第 83～85 行的 onServiceDisconnected 方法释放 upDateService 对象。

7.4.2 服务器

本节将为读者讲解与 Chapter07_Client 对应的服务器。首先运行 Chapter07_Server，查看效果图 7-14 所示。

Chapter07_Server 直接将新增加的图书发送到客户端，通知客户端查看新增图书。下面是 Chapter07_Server 的包结构，如图 7-15 所示。

与 Chapter06 的 TCP 服务器相比，这里修改了界面 Main，并在 net.cqupt 包中增加了一个 ServerThread 类，这个类新开了一个端口与客户端进行通信。首先来看一下 Main 的代码：

第 7 章 图书管理系统程序进阶

图 7-14　Chapter07_Server 效果

图 7-15　包结构

```
1.  package ui.cqupt;
2.  import java.awt.EventQueue;
3.
4.  import javax.swing.JFrame;
5.  import javax.swing.JPanel;
6.  import javax.swing.border.EmptyBorder;
7.  import javax.swing.GroupLayout;
8.  import javax.swing.GroupLayout.Alignment;
9.  import javax.swing.JTextField;
10. import javax.swing.JLabel;
11. import javax.swing.LayoutStyle.ComponentPlacement;
12. import javax.swing.JButton;
13.
14. import control.cqupt.Controller;
15. import net.cqupt.Server;
16. import net.cqupt.ServerService;
17.
18. import java.awt.Font;
19. import java.awt.event.ActionEvent;
20. import java.awt.event.ActionListener;
21. import java.io.IOException;
22. import java.io.OutputStream;
23. import java.net.Socket;
24. import java.util.HashMap;
25. import java.util.Iterator;
26. import java.util.Map;
```

```
27.    import java.util.Map.Entry;
28.    import java.util.Set;
29.
30.    public class Main extends JFrame implements ActionListener{
31.
32.        private static final long serialVersionUID = 1L;
33.        private JPanel contentPane;
34.        private JTextField idField;
35.        private JTextField nameField;
36.        private JTextField priceField;
37.        Server server;
38.        ServerService serverService;
39.
40.        public static void main(String[] args) {
41.            EventQueue.invokeLater(new Runnable() {
42.                public void run() {
43.                    try {
44.                        Main frame = new Main();
45.                        frame.setVisible(true);
46.                    } catch (Exception e) {
47.                        e.printStackTrace();
48.                    }
49.                }
50.            });
51.        }
52.
53.        public Main() {
54.            server=new Server();
55.            server.start();
56.            serverService=new ServerService();
57.            serverService.start();
58.            setTitle("server");
59.            setDefaultCloseOperation(JFrame.EXIT_ON_CLOSE);
60.            setBounds(100, 100, 450, 300);
61.            contentPane = new JPanel();
62.            contentPane.setBorder(new EmptyBorder(5, 5, 5, 5));
63.            setContentPane(contentPane);
64.
65.            idField = new JTextField();
66.            idField.setColumns(10);
67.
68.            JLabel id = new JLabel("图书编号");
69.
70.            JLabel name = new JLabel("图书名称");
```

```
71.
72.        nameField = new JTextField();
73.        nameField.setColumns(10);
74.
75.        JLabel price = new JLabel("图书价格");
76.
77.        priceField = new JTextField();
78.        priceField.setColumns(10);
79.
80.        JButton sendbutton = new JButton("发送");
81.        sendbutton.addActionListener(this);
82.
83.        JLabel title = new JLabel("新增图书");
84.        title.setFont(new Font("宋体", Font.BOLD, 20));
85.        GroupLayout gl_contentPane = new GroupLayout(contentPane);
86.        gl_contentPane.setHorizontalGroup(
87.            gl_contentPane.createParallelGroup(Alignment.LEADING)
88.                .addGroup(gl_contentPane.createSequentialGroup()
89.                    .addGroup(gl_contentPane.createParallelGroup(Alignment.LEADING)
90.                        .addGroup(gl_contentPane.createSequentialGroup()
91.                            .addGap(111)
92.                    .addGroup(gl_contentPane.createParallelGroup(Alignment.LEADING, false)
93.                        .addGroup(gl_contentPane.createSequentialGroup()
94.                        .addComponent(price)
95.                        .addPreferredGap(ComponentPlacement.UNRELATED)
96.                                .addComponent(priceField))
97.                        .addGroup(gl_contentPane.createSequentialGroup()
98.                                .addComponent(name)
99.                        .addPreferredGap(ComponentPlacement.UNRELATED)
100.                                .addComponent(nameField))
101.                        .addGroup(gl_contentPane.createSequentialGroup()
102.                                .addComponent(id)
103.                        .addPreferredGap(ComponentPlacement.UNRELATED)
104.                        .addGroup(gl_contentPane.createParallelGroup(Alignment.LEADING, false)
105.                                        .addComponent(idField)
106.                                        .addComponent(title,
GroupLayout.DEFAULT_SIZE, GroupLayout.DEFAULT_SIZE, Short.MAX_VALUE)))))
107.                        .addGroup(gl_contentPane.createSequentialGroup()
```

```
108.                            .addGap(161)
109.                            .addComponent(sendbutton)))
110.                    .addContainerGap(162, Short.MAX_VALUE))
111.            );
112.            gl_contentPane.setVerticalGroup(
113.                gl_contentPane.createParallelGroup(Alignment.LEADING)
114.                    .addGroup(gl_contentPane.createSequentialGroup()
115.                        .addGap(22)
116.                        .addComponent(title)
117.                        .addGap(39)
118.                        .addGroup(gl_contentPane.
                            createParallelGroup(Alignment.BASELINE)
119.                            .addComponent(idField, GroupLayout.
                            PREFERRED_SIZE, GroupLayout.DEFAULT_
                            SIZE, GroupLayout.PREFERRED_SIZE)
120.                            .addComponent(id))
121.                        .addPreferredGap(ComponentPlacement.
                            UNRELATED)
122.                        .addGroup(gl_contentPane.
                            createParallelGroup(Alignment.BASELINE)
123.                            .addComponent(name)
124.                            .addComponent(nameField, GroupLayout.
                            PREFERRED_SIZE, GroupLayout.DEFAULT_
                            SIZE, GroupLayout.PREFERRED_SIZE))
125.                        .addPreferredGap(ComponentPlacement.
                            UNRELATED)
126.                        .addGroup(gl_contentPane.
                            createParallelGroup(Alignment.BASELINE)
127.                            .addComponent(price)
128.                            .addComponent(priceField, GroupLayout.
                            PREFERRED_SIZE, GroupLayout.DEFAULT_
                            SIZE, GroupLayout.PREFERRED_SIZE))
129.                        .addGap(18)
130.                        .addComponent(sendbutton)
131.                        .addContainerGap(47, Short.MAX_VALUE))
132.            );
133.            contentPane.setLayout(gl_contentPane);
134.        }
135.
136.        public void actionPerformed(ActionEvent e){
137.            String bid=idField.getText();
138.            String bname=nameField.getText();
```

```
139.            String bprice=priceField.getText();
140.            idField.setText("");
141.            nameField.setText("");
142.            priceField.setText("");
143.            Controller controller=new Controller();
144.            String mes=controller.upDate(bid, bname, bprice);
145.            HashMap<String, Socket> socketPool =serverService.
                getSocketPool();
146.            Set entry = socketPool.entrySet();
147.            Iterator iter = entry.iterator();
148.            OutputStream out=null;
149.            while(iter.hasNext())
150.            {
151.                Map.Entry<String, Socket> name= (Entry<String,
                    Socket>) iter.next();
152.                Socket socket=name.getValue();
153.                try {
154.                    out=socket.getOutputStream();
155.                    out.write(mes.getBytes("GBK"));
156.                    out.flush();
157.                } catch (IOException e1) {
158.                    e1.printStackTrace();
159.                }
160.            }
161.        }
162.    }
```

Main 类定义了服务器的界面，在构造函数中完成了界面的布局和添加组件，本章节的所有组件都是通过 Eclipse 的 windowsBuilder 插件完成的。在构造函数的第 54~57 行里开启连接客户端的线程，其中第 54、55 行的 Server 线程在前面就讲过了，这里主要介绍 ServerService 线程，代码如下所示：

```
1.  package net.cqupt;
2.
3.  import java.io.IOException;
4.  import java.net.ServerSocket;
5.  import java.net.Socket;
6.  import java.util.HashMap;
7.
8.  public class ServerService extends Thread {
9.      private ServerSocket serverscoket;
10.     private HashMap<String, Socket> socketPool = new HashMap<String,
        Socket>();
```

```
11.
12.     public void run() {
13.         try {
14.             serverscoket = new ServerSocket(8003);
15.             while (!ServerService.interrupted()) {
16.                 Socket socket = serverscoket.accept();
17.                 String name = socket.getLocalAddress().
                        getHostAddress() + "::"
18.                         + socket.getPort();
19.                 socketPool.put(name, socket);
20.             }
21.         } catch (IOException e) {
22.             e.printStackTrace();
23.         } finally {
24.             close();
25.         }
26.
27.     }
28.
29.
30.     public HashMap<String, Socket> getSocketPool()
31.     {
32.      return socketPool;
33.     }
34.
35.     public void deleteThread(String name) {
36.         Socket socket = socketPool.get(name);
37.         try {
38.             socket.close();
39.             socketPool.remove(name);
40.         } catch (IOException e) {
41.             e.printStackTrace();
42.         }
43.
44.     }
45.
46.     public void close() {
47.         if (serverscoket != null) {
48.             try {
49.                 serverscoket.close();
50.             } catch (IOException e) {
51.                 e.printStackTrace();
```

```
52.            }
53.          }
54.        }
55. }
```

ServerService 类在第 10 行定义了一个 HashMap 类型的变量,用于存放连入的客户端的 Socket 对象,在 run 方法中实现了将所有连入的客户端 Socket 对象放入到 HashMap 中存储,第 30～33 行的 getSocketPool 函数返回这个成员变量。第 35～44 行的 deleteThread 方法通过得到客户端的名称,删除存储在 HashMap 中的客户端 Socket,并且中断连接。

下面回到 Main 类的第 136～161 行的事件处理函数中,第 37～42 行得到文本框中的数据,并且把文本框设置为空。第 143、144 行通过调用 Controller 中的 upDate 方法对从文本框中得到的数据进行操作,Controller 类中的 upDate 方法代码如下所示:

```
1. public String upDate(String id,String name,String price)
2. {
3.     Packager packager=new Packager();
4.     Book book=new Book(id,name,price);
5.     BookList booklist=new BookList();
6.     booklist.insert(book);
7.     String mes=packager.BookPackage(book);
8.     return mes;
9. }
```

在 upDate 方法中首先把新增加的图书插入到 BookList 和数据库中,通过自定义的协议打包,然后转换成字符串返回。

回到 Main 类的第 150～160 行是通过 Iterator 对象遍历从 serverService 对象中得到的存储客户端 Socket 的 HashMap 对象,第 154～156 行首先得到客户端的 Socket 连接,然后将数据发送给客户端,并且刷新缓冲区。这就是整个服务器的实现。

7.5 带异步刷新功能的图书管理系统

本节将为图书管理系统添加一个全新的功能——异步刷新,并为它换个界面。本节的图书管理系统是在第 3 章的多界面图书管理系统上修改的,对界面进行了美化,增加了 Tab 选项卡和"上拉更新、下拉更多"的效果。下面先展示程序的界面效果如图 7-16 所示。

下面再来看看工程的包结构,如图 7-17 所示。

由于工程是在 Chapter 3.9 基础上修改的,所以包结构很相似。这里就介绍一下新增的几个包。image.cqupt 包里中只有一个 FilesTool 类,从图 7-16 可以看到,图书列表的图书信息中包含一些图片信息,这些图片是存在数据库里的。FilesTool 中封装了获取二进制图片的方法,可以通过图书 id 获取对应的图片。我们还为程序增加了"上拉刷新、下拉更多"的效果,pull.lib.cqupt 的作用就是实现这个效果,这个效果借用了网上一个实例,有兴趣的读者可以去下载参考。网址是 http://github.com/chrisbanes/Android-

PullToRefresh。包结构就介绍到这里了，下面为大家详细讲解程序。

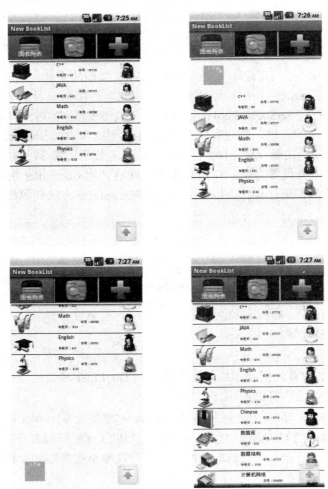

图 7-16 程序界面效果

7.5.1 Tab 标签的实现

和以往所讲的图书管理系统比较起来，本节的图书管理系统在界面上有了很大的改变，如图 7-18 所示。

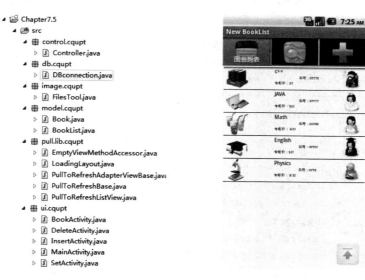

图 7-17　包结构图　　　　图 7-18　图书管理系统界面

这个图书管理系统采用了 Tab 选项卡的形式将几个主要的 Activity（BookActivity、InsertActivity、SetActivity）对应显示在图书列表、插入、编辑 3 个选项卡中。下面就按步骤来完成这些选项卡。

（1）为每个选项卡选择两张图片，作为选中状态和未选中状态的选项卡图片，并将图片存放在 res/drawable 文件夹里，如图 7-19 所示。

（2）在 res/drawable 文件夹下为每个选项卡新建 XML 文件，分别命名为 ic_tab_book、ic_tab_insert、ic_tab_set。3 个 XML 的配置类似，例如 ic_tab_book.xml 的代码如下：

图 7-19　选项卡图片

```
1.  <?xml version="1.0" encoding="utf-8"?>
2.  <selector xmlns:android="http://schemas.android.com/apk/res/android">
3.  
4.      <!-- When selected -->
5.      <item android:drawable="@drawable/ic_book_af"
6.          android:state_selected="true"/>
7.      <!-- When not selected -->
8.      <item android:drawable="@drawable/ic_book_pre"/>
9.  
10. </selector>
```

第 5、8 行的代码的作用就是分别配置选中和位选中状态的图标，其余两个 XML 仿效就行了。

（3）程序的 Tab 标签是配置在 MainActivity 中的，但在编写 MainActivity 前需先编写 MainActivity 的界面 XML 文件 main.xml。main.xml 代码如下：

```
1.  <?xml version="1.0" encoding="utf-8"?>
2.  <TabHost xmlns:android="http://schemas.android.com/apk/res/android"
3.      android:id="@android:id/tabhost"
4.      android:layout_width="fill_parent"
5.      android:layout_height="fill_parent"
6.      android:background="#FFFFFF"
7.      >
8.      <LinearLayout
9.          android:layout_width="fill_parent"
10.         android:layout_height="fill_parent"
11.         android:orientation="vertical" >
12.         <TabWidget
13.             android:id="@android:id/tabs"
14.             android:layout_width="fill_parent"
15.             android:layout_height="wrap_content"
16.             android:background="#afeeee"
17.             />
18.         <FrameLayout
19.             android:id="@android:id/tabcontent"
20.             android:layout_width="fill_parent"
21.             android:layout_height="fill_parent"
22.             android:padding="2dp" />
23.     </LinearLayout>
24. </TabHost>
```

（4）编写 MainActivity，MainActivity 的代码如下：

```
1.  import android.app.TabActivity;
2.  import android.content.Intent;
3.  import android.content.res.Resources;
4.  import android.os.Bundle;
5.  import android.widget.TabHost;
6.  import db.cqupt.DBconnection;
7.
8.  public class MainActivity extends TabActivity {
9.
10.     public void onCreate(Bundle savedInstanceState) {
11.         super.onCreate(savedInstanceState);
12.         DBconnection.setContext(this.getApplicationContext());
```

```
13.        setContentView(R.layout.main);
14.        Resources res = getResources();
15.        TabHost tabHost = getTabHost();
16.        TabHost.TabSpec spec;
17.        Intent intent;
18.        intent = new Intent().setClass(this, BookActivity.class);
19.        spec = tabHost.newTabSpec("allBooks")
20.            .setIndicator("图书列表", res.getDrawable(R.drawable.
               ic_tab_book))
21.            .setContent(intent);
22.        tabHost.addTab(spec);
23.
24.        intent = new Intent().setClass(this, InsertActivity.class);
25.        spec = tabHost.newTabSpec("insert")
26.            .setIndicator("插入", res.getDrawable(R.drawable.
               ic_tab_insert))
27.            .setContent(intent);
28.        tabHost.addTab(spec);
29.
30.        intent = new Intent().setClass(this, SetActivity.class);
31.        spec = tabHost.newTabSpec("set")
32.            .setIndicator("编辑", res.getDrawable
               (R.drawable.ic_tab_set))
33.            .setContent(intent);
34.        tabHost.addTab(spec);
35.
36.        tabHost.setCurrentTab(0);
37.    }
38. }
```

首先，MainActivity 要继承 TabActivity，然后主要的代码在 onCreate 函数里，第 18～34 行，设置 Tab 的每个选项卡，如第 18～22 行，第 18 行为选项卡创建一个 Intent，从而能完成选项卡的跳转，因为选项卡的跳转其实就是 Activity 的切换。setIndicator 是在选项卡设置文字和图标。最后第 36 行是为界面设置一个默认选中的选项卡。

随着技术的不断发展，TabActivity 已经不建议使用了，新的技术要求我们使用 Fragments 代替 TabActivity，但这里不再讲解 Fragments 的用法，读者可以将它作为课后作业来研究。

7.5.2 自定义的 ListView 与 Adapter

本节的图书管理系统在界面上做了美化，之前所用到的 ListView 的每个 Item 只是简单的一个组件，这里的图书管理系统的 ListView 的每个 Item 是由多种组件组成的，如图 7-20 所示。这些 Item 的布局文件 items.xml 文件代码如下：

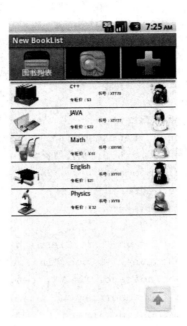

图 7-20　ListView

```
1.  <?xml version="1.0" encoding="utf-8"?>
2.  <RelativeLayout xmlns:android="http://schemas.android.com/apk/res/android"
3.      android:layout_width="fill_parent"
4.      android:layout_height="fill_parent"
5.      android:background="#FFFFFF"
6.      android:orientation="horizontal" >
7.
8.      <ImageView
9.          android:id="@+id/imageView_item_book"
10.         android:layout_width="60dip"
11.         android:layout_height="50dip"
12.         android:src="@+drawable/shu_2" />
13.
14.     <TextView
15.         android:id="@+id/name_item"
16.         android:layout_width="220dip"
17.         android:layout_height="wrap_content"
18.         android:layout_alignParentTop="true"
19.         android:layout_alignRight="@+id/price_item"
20.         android:layout_toRightOf="@+id/imageView_item_book"
21.         android:textColor="#000000"
22.         android:textSize="12dp" />
23.
24.     <TextView
```

```
25.        android:id="@+id/id_item"
26.        android:layout_width="220dip"
27.        android:layout_height="wrap_content"
28.        android:layout_alignLeft="@+id/name_item"
29.        android:layout_below="@+id/name_item"
30.        android:layout_toLeftOf="@+id/imageView_item_author"
31.        android:textColor="#000000"
32.        android:textSize="8dp" />
33.
34.    <TextView
35.        android:id="@+id/price_item"
36.        android:layout_width="220dip"
37.        android:layout_height="wrap_content"
38.        android:layout_below="@+id/id_item"
39.        android:layout_toLeftOf="@+id/imageView_item_author"
40.        android:layout_toRightOf="@+id/imageView_item_book"
41.        android:textColor="#000000"
42.        android:textSize="8dp" />
43.
44.    <ImageView
45.        android:id="@+id/imageView_item_author"
46.        android:layout_width="60dip"
47.        android:layout_height="50dip"
48.        android:layout_alignParentRight="true"
49.        android:layout_alignParentTop="true"
50.        android:src="@+drawable/author_1" />
51. </RelativeLayout>
```

可以看到，ListView 的每个 Item 由多个 TextView 和 ImageView 组成。既然 ListView 变复杂了，那么以前用的那些 Adapter 就不能满足这个 ListView 了，于是我们为这个 ListView 量身打造了一个 Adapter——RatingAdapter，代码如下：

```
1.  private class RatingAdapter extends BaseAdapter {
2.      private Context context;
3.      LayoutInflater layoutInflater;
4.      String inflater = Context.LAYOUT_INFLATER_SERVICE;
5.      public RatingAdapter(Context context) {
6.          this.context = context;
7.          layoutInflater = (LayoutInflater) this.context
8.                  .getSystemService(inflater);
9.      }
10.     public View getView(int position, View convertView, ViewGroup
            parent) {
11.         Book book = booklist.get(position);
```

```
12.            String id = book.getId();
13.            String name = book.getName();
14.            String price = book.getPrice();
15.            RelativeLayout item = (RelativeLayout) layoutInflater.inflate
               (R.layout.items, null);
16.            ImageView bookImage = (ImageView) item.findViewById
               (R.id.imageView_item_book);
17.            ImageView authorImage = (ImageView) item.findViewById
               (R.id.imageView_item_author);
18.            FilesTool fileReader = new FilesTool(id.trim());
19.            authorImage.setImageBitmap(fileReader.getAuthorImage());
20.            bookImage.setImageBitmap(fileReader.getBookImage());
21.            TextView nameT = (TextView) item.findViewById
               (R.id.name_item);
22.            TextView idT = (TextView) item.findViewById(R.id.id_item);
23.            TextView priceT = (TextView) item.findViewById
               (R.id.price_item);
24.            nameT.setText("      " + name);
25.            idT.setText("      书号: " + id);
26.            priceT.setText("      专柜价: " + price);
27.            return item;
28.        }
29.        public int getCount() {
30.            return booklist.size();
31.        }
32.        public Object getItem(int position) {
33.            return booklist.get(position);
34.        }
35.        public long getItemId(int position) {
36.            return position;
37.        }
38.    }
```

自定义的 RatingAdapter 继承了 BaseAdapter，并实现了其中重要的几个方法，构造函数中的 Context 即当前的 Activity，用于获取 LayoutInflater 对象，该对象的作用类似于函数 findViewById()，用于查找 layout 文件夹下的 XML 布局文件，该对象在 getView() 方法中有重要作用，这里主要讲解 getView(int position, View convertView, ViewGroup parent)方法。该方法的作用就是获取 ListView 中 position 位置的 View，有了 position 参数，就可以在图书列表中找到对应位置的图书信息，从而达到显示每本书信息的目的。在该函数中返回自定义的 View 作为该位置的 Item，所以 getView 方法是 Adapter 的核心。

通过第 15 行代码可得到一个 Layout 对象，该对象通过 layoutInflater 找到的 items.xml 对象，代码第 16~26 行分别获得 items.xml 中的各个组件并设置其属性，最后返回设置完成的 Layout 对象，最后界面上显示的就是该位置图书的信息了。

7.5.3 异步刷新实现

这里说的刷新包括程序中的刷新和更多功能，两者的实现机制是差不多的。功能如图 7-21 所示。

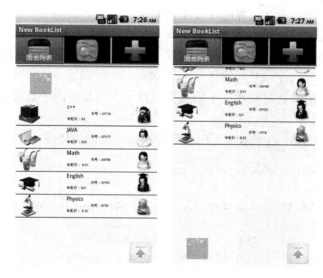

图 7-21 上拉刷新、下拉更多

在第 3 章中我们讲过，Android 界面的更新必须由主线程来完成，这里的界面刷新就涉及这个问题，当遇到这种问题时可以采用 Handler 来解决，但这样的操作过于烦琐，所以采用了 Android 中的一个非常重要的类——AsyncTask。它用于帮助执行程序中开销较大操作。

GetDataTask 类继承了 AsyncTask 类，并重载了 AsyncTask 的 doInBackground()方法和 onPostExecute()方法，但在讲解这些方法前需要讲一下 AsyncTask 后面的 3 种泛型类型：Params、Progress 和 Result。

- Params：发送给任务执行时的参数的类型。
- Progress：后台任务执行的百分比。
- Result：后台执行任务最终返回的结果。

AsyncTask 的执行共分为 4 个步骤，每个步骤都有一个回调方法，这些方法都由系统调用，而我们要做的就是将 AsyncTask 子类化，并实现下面的一个或几个方法。

- onPreExecute()：在任务被调用后该方法会被立即调用，在该方法中可以做一些准备工作，如显示一个进度条。
- doInBackground(Params...)：抽象方法，子类必须实现。在 onPreExecute 方法执行后立即调用，该方法被放在后台线程中执行，主要用于执行那些比较耗时的操作。比如，我们就将图书列表的后台数据的更新工作放在该方法中完成。在这一步中，还可以调用 publishProgress(Progress...)方法来更新实时的任务进度。
- onProgressUpdate(Progress...)：在 publishProgress(Progress...)被调用后，UI 线程将

调用该方法更新任务的进行情况。
- onPostExecute(Result)：在 doInBackground 执行完成后，该方法将被 UI 线程调用，后台的计算结果将通过该方法传递到 UI 线程。.

除此之外，使用 AsyncTask 还需注意以下 4 点：
（1）该类的子类必须在主界面线程中被创建。
（2）通过 execute()方法启动该任务。
（3）不能手动启动 onPreExecute()、doInBackground()、onProgressUpdate()、onPostExecute()方法。
（4）该任务只能被执行一次，不能重复执行，也就是说，子类对象的 execute()方法只能被调用一次。

了解完 AsyncTask，再来看一下图书管理系统中刷新的代码。代码如下：

```
1.   private class GetDataTask extends AsyncTask<Void, Void, ArrayList
        <Book>> {
2.         protected ArrayList<Book> doInBackground(Void... params) {
3.             try {
4.                 Thread.sleep(2000);
5.             } catch (InterruptedException e) {
6.             }
7.             booklist.clear();
8.             Controller control = new Controller();
9.             BookList myBookList = control.searchBook();
10.            for (int i = 0; i < 5 && i < myBookList.size(); ++i) {
11.                booklist.add(myBookList.get(i));
12.            }
13.            return booklist;
14.        }
15.        protected void onPostExecute(ArrayList<Book> result) {
16.            mPullRefreshListView.onRefreshComplete();
17.            raAdapter.notifyDataSetInvalidated();
18.            super.onPostExecute(result);
19.        }
20.    }
```

在这个版本的图书管理系统中只实现了 AsyncTask 的两个方法：doInBackground、onPostExecute。在 doInBackground 中将界面的的图书列表清空并重新加载了书库中的前 5 本书，这样就完成了图书列表的后台数据更新。在 onPostExecute 方法中完成了界面的更新，notifyDataSetInvalidated 方法的作用就是重新绘制 ListView 控件。

7.5.4 其他部分实现

本节所讲的图书管理系统采用的是单机版，所有的图书信息都是预先存在 SQLite 数据库中随程序一起发布的，图 7-22 就是数据的表结构，图 7-23 是工程的 res 文件夹，其中 raw 中存放的就是该数据库。

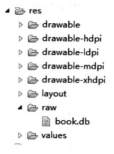

图 7-22　数据库结构　　　　　　图 7-23　res 文件夹

在第一次调用 DBconnection 类连接数据库时，DBconnection 会调用它的 openDatabase 方法，将 book.db 文件存放到 SD 卡中。DBconnection 代码如下：

```
1.  import java.io.File;
2.  import java.io.FileNotFoundException;
3.  import java.io.FileOutputStream;
4.  import java.io.IOException;
5.  import java.io.InputStream;
6.
7.  import ui.cqupt.R;
8.  import android.content.Context;
9.  import android.database.Cursor;
10. import android.database.sqlite.SQLiteDatabase;
11.
12. public class DBconnection {
13.     private final static String DATABASE_PATH = "/sdcard/books/";
14.     private final static String DATABASE_FILENAME = "book.db";
15.
16.     private static Context context;
17.
18.     public static void setContext(Context applicationContext) {
19.         context = applicationContext;
20.     }
21.
22.     public DBconnection() {
23.
24.     }
25.
26.     public void onUpgrade(SQLiteDatabase db, int oldVersion, int
        newVersion) {
27.
28.     }
```

```
29.
30.    public SQLiteDatabase getConnection() {
31.
32.        return openDatabase();
33.    }
34.
35.    public void close(SQLiteDatabase db,Cursor cur) {
36.        if(cur != null)
37.            cur.close();
38.        if (db != null)
39.            db.close();
40.    }
41.
42.    private SQLiteDatabase openDatabase() {
43.        try {
44.            String databaseFilename = DATABASE_PATH + DATABASE_FILENAME;
45.            File dir = new File(DATABASE_PATH);
46.            if (!dir.exists()) {
47.                dir.mkdir();
48.            }
49.            if (!(new File(databaseFilename)).exists()) {
50.                InputStream is = context.getResources().openRawResource(
51.                        R.raw.book);
52.                FileOutputStream fos = new FileOutputStream(databaseFilename);
53.                byte[] buffer = new byte[8192];
54.                int count = 0;
55.                while ((count = is.read(buffer)) > 0) {
56.                    fos.write(buffer, 0, count);
57.                }
58.
59.                fos.close();
60.                is.close();
61.            }
62.            SQLiteDatabase database = SQLiteDatabase.openOrCreateDatabase(
63.                    databaseFilename, null);
64.            return database;
65.        } catch (FileNotFoundException e) {
66.            e.printStackTrace();
67.        } catch (IOException e) {
68.            e.printStackTrace();
```

```
69.            }
70.            return null;
71.        }
72. }
```

还有就是之前提到过的 image.cqupt 包中的 FilesTool 类，它用于在数据库中读取图片文件，代码如下：

```
1.  import android.database.Cursor;
2.  import android.database.sqlite.SQLiteDatabase;
3.  import android.graphics.Bitmap;
4.  import android.graphics.BitmapFactory;
5.  import db.cqupt.DBconnection;
6.
7.  public class FilesTool {
8.
9.      private Bitmap bookBitmap = null;
10.     private Bitmap authorBitmap = null;
11.
12.
13.     public FilesTool(String id) {
14.         readImage(id);
15.     }
16.
17.     public Bitmap getAuthorImage() {
18.
19.         if (authorBitmap == null)
20.             readImage("delault");
21.         return authorBitmap;
22.     }
23.
24.     public Bitmap getBookImage() {
25.         if (bookBitmap == null)
26.             readImage("delault");
27.         return bookBitmap;
28.     }
29.     private void readImage(String id) {
30.         DBconnection connection = new DBconnection();
31.         SQLiteDatabase db = connection.getConnection();
32.         Cursor cur = null;
33.
34.         cur = db.query("image", new String[] { "bookicon", "author" },
35.                 "id= ? ", new String[] { id }, null, null, null);
36.         int bookIndex = cur.getColumnIndex("bookicon");
```

```
37.         int authorIndex = cur.getColumnIndex("author");
38.
39.         while (cur.moveToNext()) {
40.             byte[] bookicon = null;
41.             byte[] authoricon = null;
42.             bookicon = cur.getBlob(bookIndex);
43.             authoricon = cur.getBlob(authorIndex);
44.             bookBitmap = BitmapFactory.decodeByteArray(bookicon, 0,
45.                     bookicon.length);
46.             authorBitmap = BitmapFactory.decodeByteArray(authoricon, 0,
47.                     authoricon.length);
48.         }
49.
50.         connection.close(db, cur);
51.     }
52. }
```

readImage 方法的作用就是从数据库中获取作者图片和书的图片。图片是存在 image 表中的，每本书的作者和图标都是以 bookid 为主键的。如果这本书没有设置图标就以默认图标作为图标。代码第 42～47 行就是从数据库中读取图片信息并以 Bitmap 的形式返回给调用者，最后显示在界面上。

以上就是异步刷新的图书管理界面的主要功能讲解。有了以上了解，读者就可以查看本书配套资源中的工程 Chapter7.5，做进一步研究。

参 考 文 献

[1] 于国防，徐永刚，张玉杰. Android 应用程序开发教程[M]. 北京：清华大学出版社，2017.
[2] 刘贤锋. Android 应用开发项目实战[M]. 北京：机械工业出版社，2017.
[3] 明日学院. Android 开发从入门到精通（项目案例版）[M]. 北京：水利水电出版社，2017.
[4] 杨谊. Android 移动应用开发[M]. 北京：人民邮电出版社，2017.
[5] 张亚运. Android 开发入门百战经典[M]. 北京：清华大学出版社，2017.
[6] 雷擎，伊凡. 基于 Android 平台的移动互联网应用开发[M]. 2 版. 北京：清华大学出版社，2017.
[7] (美)Bill Phillips，Chris Stewart，Kristin Marsicano. Android 编程权威指南[M]. 3 版. 王明发，译. 北京：人民邮电出版社，2017.
[8] (美)Phil Dutson. Android 开发模式与最佳实践[M]. 李雄，译. 北京：电子工业出版社，2017.
[9] 毕小明. 精通 Android Studio[M]. 北京：清华大学出版社，2017.